OP AMPS

Made Easy

Concepts

Troubleshooting

OP AMPS Made Easy

Concepts and Troubleshooting

JOHN EDWARD HART
DeVry Institute of Technology

Prentice Hall, *Englewood Cliffs, New Jersey 07632*

Library of Congress Cataloging-in-Publication Data

Hart, John Edward.
Op-amps made easy: concepts and troubleshooting / John Edward Hart.
p. cm.
Bibliography: p.
Includes index.
ISBN 0-13-638321-1
1. Operational amplifiers. I. Title.
TK7871.58.O6H37 1989
621.3815'35—dc 19

Editorial/production supervision and
interior design: Lillian Glennon
Cover design: 20/20 Services Inc.
Manufacturing buyer: Robert Anderson

A Division of Simon & Schuster
Englewood Cliffs, New Jersey 07632

Printed in the United States of America
10 9 8 7 6 5 4 3 2 1

ISBN 0-13-638321-1

Prentice-Hall International (UK) Limited, *Londo*
Prentice-Hall of Australia Pty. Limited, *Sydney*
Prentice-Hall Canada Inc., *Toronto*
Prentice-Hall Hispanoamericana, S.A., *Mexico*
Prentice-Hall of India Private Limited, *New Delhi*
Prentice-Hall of Japan, Inc., *Tokyo*
Simon & Schuster Asia Pte. Ltd., *Singapore*
Editora Prentice-Hall do Brasil, Ltda., *Rio de Janeiro*

Contents

Preface

Discussions of basic operational amplifier properties can be found in many electronic texts covering modern solid-state components. However, while most of them cover the operational amplifier's own properties quite adequately, the variety of operational amplifier applications discussed is often limited to only the simplest and most common circuit formats.

A limited number of texts devoted to the operational amplifier are already in existence. While these texts are very thorough in coverage of operational amplifier applications, the advanced mathematics used to analyze and describe circuit behavior, necessary for the degree-level student, can prove to be quite unfathomable for other students with a less-detailed mathematical background.

I have written this text to provide the student of electronic devices with a thorough coverage of both the properties and the applications of the operational amplifier, while keeping necessary mathematics as simple as possible, involving only basic algebra techniques in the development and utilization of all governing formulas.

John E. Hart

Acknowledgements

I want to thank everyone who helped me to write this book. First, I would like to thank my editor, Alice Barr, for having the confidence in me to convince me to write the book! Lillian Glennon and Mariann Hutlak have provided very professional support in spearheading production of the book. I would also like to thank Les Taylor and David M. Newman who reviewed the original manuscript. I must thank Dean Eugene Miller for compassionate scheduling during my writing ordeal. Finally, I owe thanks to my parents and friends for understanding my times of seclusion since the book's beginnings.

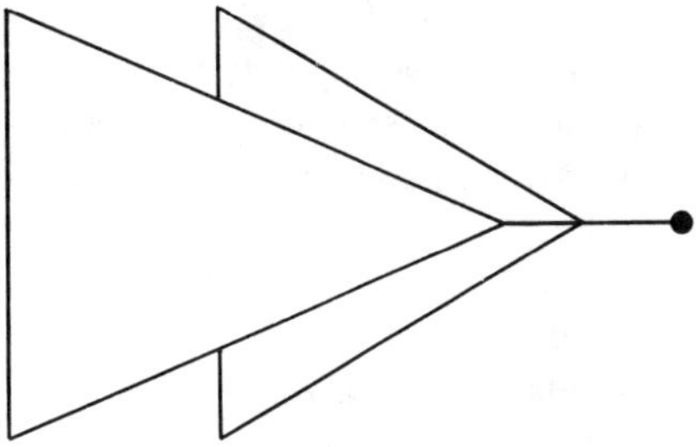

Introduction

The **operational amplifier,** or ***op amp*** for short, was originally developed for use in specialized circuits which, through their electrical response characteristics, closely approximate certain useful mathematical operations. Since it was a fairly complex circuit built of separate or discrete components, the early op amp was fairly bulky and expensive. However, since its basic design format was that of a high-gain multistage amplifier, further applications for the op amp were soon developed, increasing the demand for readily available op amp circuitry. Production of the op amp in integrated-circuit form made this versatile electronic tool available to the public at a very reasonable cost and in an array of case styles, output capabilities, and ranges of operating frequency.

Part of the appeal of the op amp lies in its simplicity of use compared with discrete transistors. Both bipolar-junction transistors and field-effect transistors require fairly elaborate circuits around them to insure proper bias conditions. Transistors also generally require input and output signal-coupling devices, such as transformers or large capacitors, owing to the necessary DC voltage levels which, although required for correct biasing, must also be isolated from input and output signal connections. The presence of these coupling devices always places limitations on the low-frequency response characteristics of circuits in which they are used. In addition, although any transistor can be connected in one of three different ways, each with its own impedance, gain, and phase characteristics, each mode offers only one input terminal, along with one output terminal. This means that each transistor format can provide only one specific, fixed-phase relationship between its input and output terminals.

The op amp offers several distinct advantages in comparison to transistors. Its biasing circuitry is incorporated into the integrated circuit itself, eliminating the need for external biasing components. Since it is designed to operate from equal positive and negative DC voltage sources, the op amp can be easily used without the need for signal-coupling devices, or direct-coupled. When direct coupling is used, the op amp will exhibit no rolloff of low-frequency response, so that even DC inputs can be amplified. Finally, while the op amp has but one output terminal, its internal construction allows it to offer *two* available inputs, each with a different phase relationship between itself and the output. With such advantages over transistors, it is easy to understand why the op amp encounters larger sales and larger varieties of applications every year!

When an op amp appears in a circuit, it is almost invariably associated with a small number of passive components, usually resistors and capacitors. However, these components do not participate in the determination of bias conditions for the op amp: their sole purpose is to determine which of its wide variety of operational capabilities the op amp will provide. In this book, we shall begin by investigating the important electrical characteristics of the op amp itself. In the remainder of the book, we shall examine the diversity of electrical response properties which the op amp, in reacting to the addition of external circuit connections, can be tailored to provide.

To appreciate the basic operational parameters of the op amp, the reader must possess previous knowledge of bipolar-junction transistors. I must assume that the reader is already familiar with basic transistor biasing requirements, standard biasing circuits, and characteristics of common-emitter, common-base, and common-collector formats with regard to voltage gain, current gain, phase relationship, and input and output impedances. Since transistors are the internal components within the circuitry of the op amp which are responsible for its ability to amplify, understanding the construction, biasing, and signal-response properties of the op amp would be difficult indeed without previous transistor knowledge.

1

The Differential Amplifier

The **op amp,** unlike any single-transistor amplifier, provides *two* available inputs, each possessing properties uniquely its own. The capability of providing two separate inputs, but only one output terminal, arises from a circuit format called the **differential amplifier,** the properties of which we shall investigate in this chapter.

1.1 THE DIFFERENTIAL AMPLIFIER CIRCUIT

As depicted in Figure 1.1, the differential amplifier consists of two bipolar-junction transistors connected into biasing circuitry, which, like any other bias circuit, sets each transistor at DC levels necessary to forward-bias each base-emitter junction and reverse-bias each collector-base junction. Thus, each BJT has the same DC conditions that it normally receives from other types of bias circuitry.

Notice that the amplifier's two input terminals are associated with the base terminals of the two transistors, while the output connections are situated at the two collector terminals, between which the load is connected. Thus, each transistor has been arranged in the common-emitter format. Since it is the only transistor format which uniquely provides both voltage gain and current gain,

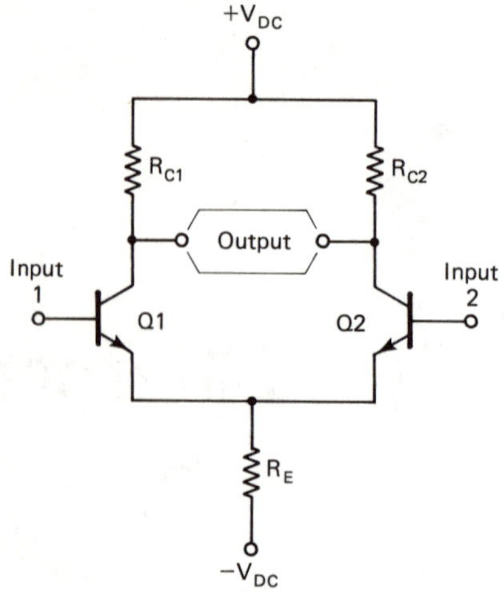

Figure 1.1

the use of the common-emitter format in the differential amplifier enables it to provide similar large gain characteristics.

Two properties of the differential amplifier arise directly from the presence of two transistors. In a single-transistor amplifier, the output point connects to one end of the load, while the other end of the load connects either to ground or to the DC supply. In the differential amplifier, however, the output is connected between the collectors of both transistors. Since both transistors will be biased similarly, both collector terminals will be at approximately the same DC potential, so that the difference in voltage between them will be quite close to zero volts. This allows the output device to be direct-coupled between the two collectors, without the need for DC isolation or the loss of gain at low frequencies that is normally associated with the use of coupling capacitors or transformers.

The second special feature of the differential amplifier is the presence of a *single* emitter resistor, through which the emitter currents of both transistors pass. The combination of both currents through the single resistor provides both transistors with exactly the same emitter voltage, setting their bias conditions to identical values. As a result, the voltage drop across the emitter resistor will affect the bias levels of both transistors. This feature becomes very important when the differential amplifier is used to amplify an AC signal supplied to either one of the two available inputs.

1.2 RESPONSE TO ONE INPUT SIGNAL

In Figure 1.2, an AC signal source has been connected to the input terminal of the differential amplifier associated with Q1. At the same time, Q2's input terminal has been connected to ground. If each transistor had its own separate

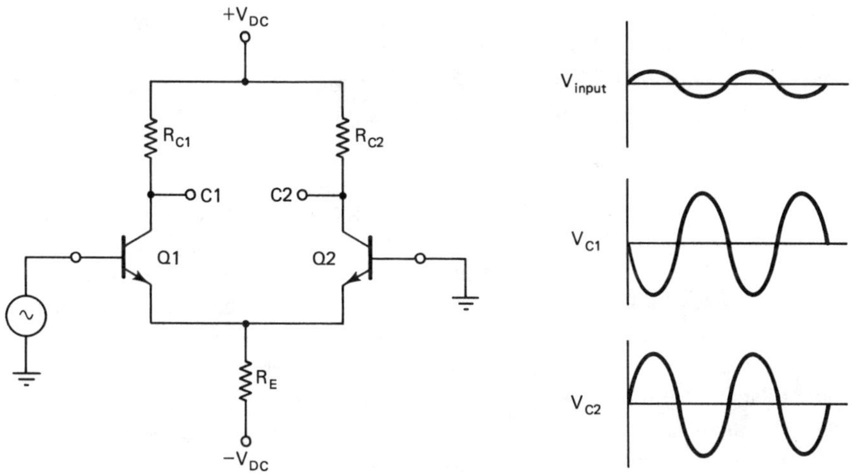

Figure 1.2

bias circuitry, Q1 would provide a variation in its collector current owing to the input signal, but Q2, receiving no input signal, would remain at its normal DC bias point. However, owing to the presence of the single emitter resistor, Q2 will also be affected by the input signal, even though it is not directly connected to Q2 at all.

When the input signal goes positive, Q1's base terminal will rise in voltage. Since the base-emitter junction is always forward-biased, its voltage drop will stay fairly constant. This results in a rise in emitter voltage along with the rise in base voltage. Q2, with its base grounded, will react to this variation in its emitter voltage as a common-base circuit, while Q1 reacts to the signal as a common emitter.

An increase in Q1's base voltage will cause an increase in base current, with a corresponding increase in collector current. This will cause an increased voltage drop across Q1's collector resistor, which causes Q1's collector voltage to decrease. At the same time, Q2, functioning as a common base, will receive less forward bias across its base-emitter junction, resulting in a decrease of base current and a corresponding decrease in collector current. This will cause a decreased voltage drop across Q2's collector resistor, which will cause Q2's collector voltage to rise. Thus, an input at Q1 will cause not only a variation in collector current for Q1 but also an *opposite* variation in collector current for Q2.

When the input signal enters a negative alternation, exactly the opposite situation will occur. A negative input signal will serve to decrease the forward bias across Q1's base-emitter junction and pull the base voltage lower. Since the base-emitter junction voltage drop will remain fairly constant, this will cause the emitter voltage to fall. The resulting decrease in base current will cause a decrease in collector and emitter current from Q1. This will cause a smaller voltage drop across Q1's collector resistor, so that Q1's collector voltage will rise. At the same time, the drop in emitter voltage serves to increase the forward bias across Q2's base-emitter junction, causing an increase in base current and a corresponding increase in collector current. The voltage drop across Q2's collector resistor will therefore increase, causing Q2's collector voltage to fall.

Note that with either a positive or a negative alternation, one situation occurred consistently: the collectors of the two transistors went in *opposite* directions. When Q1's collector rose in voltage, Q2's collector fell. When Q1's collector voltage fell, Q2's collector voltage rose. Since the output is connected between the two collectors, then the total variation between them is effectively twice the size of the variation found at one single collector terminal. When we remember that the common emitter circuit provides a 180-degree phase shift between input and output, while the common base provides an output in phase with its input, we have further evidence supporting the opposite changes which occur between the two collectors. Notice that the variation at Q1's collector is opposite, or 180 degrees out of phase, with the input signal, while Q2's collector varies in phase with the input signal.

If we reverse the input situation to that shown in Figure 1.3, connecting the input source to Q2's base, and grounding Q1's base, a similar situation will occur, except Q2 will now operate as a common emitter, and Q1, with its base grounded, will operate as a common base. Q2's collector will provide a variation 180 degrees out of phase with the input, while Q1's collector will now vary in phase with the input. Now, however, when the input source goes positive, Q1's collector will go higher while Q2's collector will go lower.

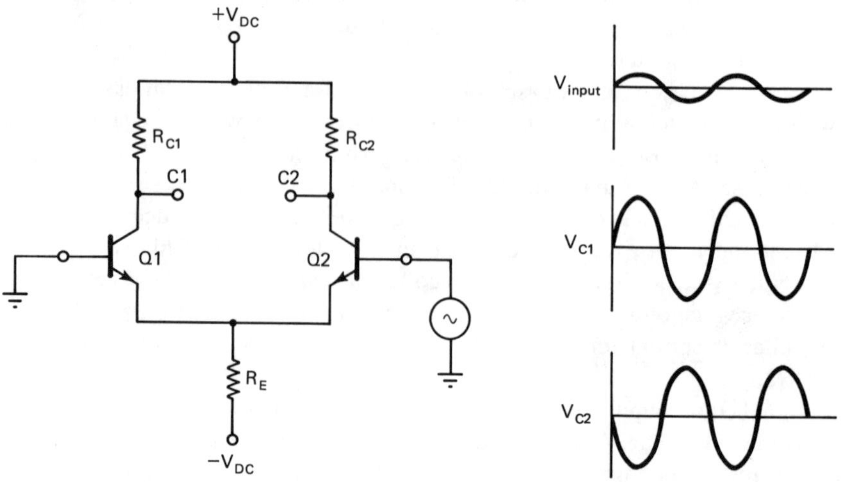

Figure 1.3

This is exactly opposite from the situation we first encountered, with the input connected to Q1. This means that if we always view the polarity of the output in the same way, one input will deliver an output signal which is *in phase* with that input, while the other input will deliver an output which is *180 degrees out of phase* with that input. We can, therefore, achieve either an in-phase output or a 180-degree-out-of-phase output, depending only on selection of the appropriate input terminal. We have also seen that, by virtue of the two transistors operating in opposite directions, the output signal developed by the differential amplifier offers up to twice the output amplitude (and, therefore, a higher gain) than a single-transistor amplifier.

1.3 RESPONSE TO TWO DIFFERENTIAL INPUTS

In the connection modes we have already discussed, one signal source has been connected to one of the two inputs, and the other unused input has been connected to ground. It is also possible, however, to connect two separate signal sources to the differential amplifier, one to each input. How the differential amplifier responds to this situation depends primarily on the phase relationship between the two input sources. The differential amplifier will respond in completely different ways to two in-phase and two out-of-phase input signals.

In Figure 1.4, the differential amplifier has one source connected to each input. Both sources deliver the same amplitude and are operating at the same AC frequency, and in this case the two sources are 180 degrees out of phase with each other, so that while one source delivers a positive alternation to its input, the other source delivers its negative alternation. In this situation we are providing **differential inputs.** Since each transistor now receives an input signal directly to its base, both transistors will now function as common emitters.

With a single input, remember that the transistor to which the signal is connected will operate as a common emitter, providing a collector output signal 180 degrees out of phase with the input. The other transistor will operate as a common base and provide a collector output signal which is in phase with the input signal. We are connecting both sources so that they provide opposite polarities at all times, characteristic of a 180-degree phase difference.

Consider the effects of source 1, connected to Q1, as it provides a positive alternation. Q1, operating common-emitter, will provide a negative alternation at its collector. Q2, operating common-base, will provide a positive alternation at its collector. Now, consider the effects of source 2, connected to Q2, as it provides a negative alternation, 180 degrees out of phase with source 1. Q2 will respond to this source as a common emitter, providing a positive alternation at its collector. Q1 will respond to this source as a common base, providing a negative alternation at its collector.

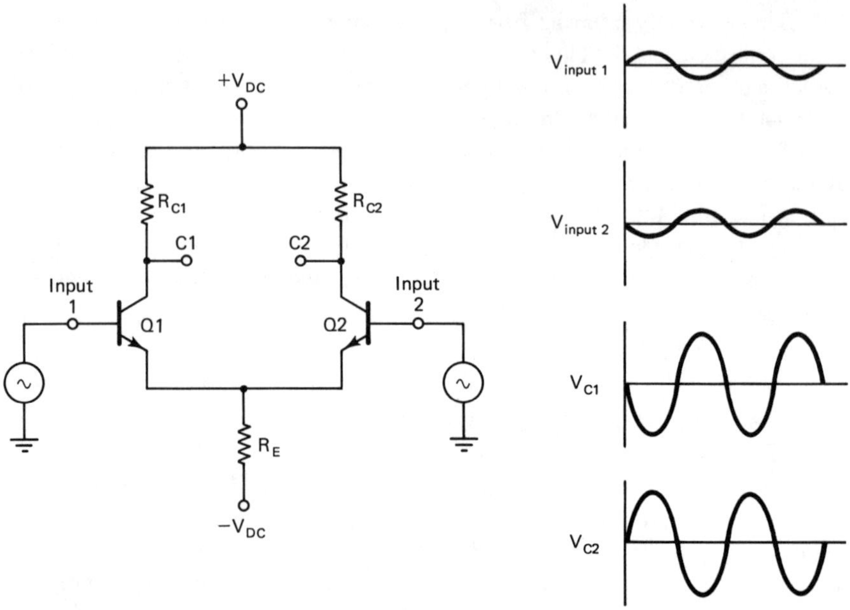

Figure 1.4

Notice that both sources drive Q1's collector lower, while at the same time they both drive Q2's collector higher. Each source would cause these effects alone; when we provide both inputs, both will work together to drive Q1's collector even lower and Q2's collector even higher. Since both cause the same type of variation, the total output variation will be approximately twice the amplitude caused by each source alone. Therefore, two differential inputs seem to provide a larger output signal than would a single input of identical size. If we consider, however, that two differential inputs will cause twice the voltage difference between the base terminals as the voltage difference seen between one input source and ground, then we must realize that the gain of the amplifier is really no larger; the output amplitude is larger because the two inputs will develop twice the voltage difference between them as that achieved by a single source. A doubled output amplitude has occurred simply because we have provided double the input amplitude.

The ability of the differential amplifier to provide a strong response to two out-of-phase input signals is, even considered by itself, an extremely significant property of the differential amplifier. This action becomes doubly significant, however, when it is compared with the differential amplifier's minimal response to the application of two input signals which are in phase with each other, rather than out of phase.

1.4 RESPONSE TO TWO IDENTICAL INPUTS

Let us now consider the response of the differential amplifier to two inputs which are equal in amplitude and *in phase* with each other. As depicted in Figure 1.5, let us connect both sources in phase to provide positive alternations simultaneously. A positive alternation at the base of Q1 will cause a negative alternation at Q1's collector and a positive alternation at Q2's collector. Now, a positive alternation at the base of Q2 will cause a negative alternation at Q2's collector and a positive alternation at Q1's collector. If we combine both results, Q1's collector is being pulled lower by its own input while simultaneously being pulled higher by Q2's input.

If the sources are equal in amplitude, the two opposing forces will balance each other, so that they cancel. This implies that the collectors will not change voltage at all! In reality, the collectors may vary somewhat, but they will both vary in the same direction, with both rising or both falling simultaneously. Since our output appears as the difference between the collectors, if they both change in the same direction, the difference between them will still be virtually zero! If we apply two equal input signals which are in phase with each other, we should normally expect little or no output signal.

This mode of operation is called **common-mode** operation. Obviously, since the gain of the differential amplifier is extremely small when operated

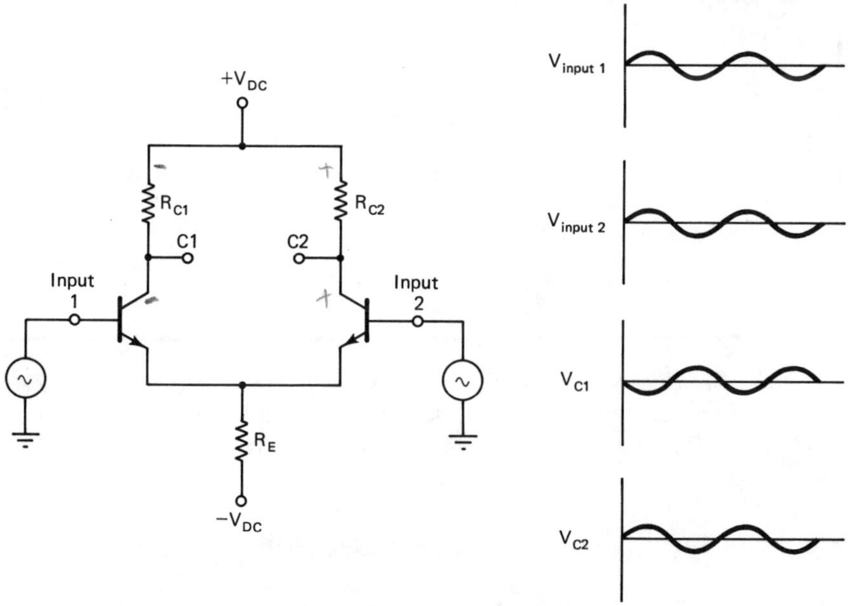

Figure 1.5

common-mode, this finding further supports the fact that the differential amplifier responds to differences between the inputs, and if the inputs are the same, little or no output will occur. Ideally, no response whatsoever should occur when common-mode inputs are provided. In reality, op amps do respond to a very limited degree to common-mode inputs.

One important op amp specification is, in fact, a measure of the op amp's ability to reject common-mode inputs. This quantity, called **common-mode rejection ratio,** or ***CMRR*** for short, is found by determining the gain figure of the op amp when responding to differential inputs. A_{diff}, and also the gain figure of the op amp when responding to common-mode inputs, A_{cm}. The CMRR figure can then be found through:

$$CMRR = \frac{A_{\text{diff}}}{A_{\text{cm}}}$$

EXAMPLE 1.1

Let us assume that we need to find the *CMRR* for a certain op amp which receives two 50-microvolt input signals. When they are in phase, the output develops a 20-microvolt amplitude. When the inputs are opposite in phase, the output develops a 10-volt amplitude.

First, we must find each gain figure. Occurring when inputs are in phase, the common-mode gain A_{cm} will be:

$$A_{\text{cm}} = \frac{20\ \mu\text{V}}{50\ \mu\text{V}} = .4$$

For out-of-phase inputs, the differential gain A_{diff} will be:

$$A_{\text{diff}} = \frac{10\ \text{V}}{50\ \mu\text{V}} = 200{,}000$$

The *CMRR* will now be:

$$CMRR = \frac{200{,}000}{.4} = 500{,}000$$

The *CMRR* should be as large a number as possible, which also suggests that the differential mode gain should be as large as possible, while the common-mode gain should be as small as possible. We shall discuss the *CMRR* further in the next chapter.

1.5 IMPROVING COMMON-MODE REJECTION

In the differential amp circuits shown in Figures 1.1 through 1.4, both transistors share a common emitter resistor. Therefore, the current through this resistor represents the sum of both transistors' emitter currents at all times. When applying common-mode input signals, positive alternations at both inputs will raise both base terminal voltages, pulling the emitters higher in voltage along with the bases. This action results in a larger voltage drop across the emitter resistor, allowing the sum of both transistors' emitter currents to increase. Since the collector currents will also increase, both collector resistors will develop larger voltage drops, causing both collector voltages to decrease. Negative alternations applied to both inputs will cause exactly opposite effects, resulting in the rise of both collector voltages.

If both transistors were identical in gain characteristics, both collectors would change by identical amounts. Since it is far more likely, however, that the transistors will have somewhat different characteristics, their collectors will actually change by slightly different amounts. Since the output is developed between the collector terminals, if both change by the same identical amount, no output would appear. Typically, however, since the collectors will change by slightly different amounts, a small common-mode output signal will normally be seen.

When operating in the differential mode, the sum of the two emitter currents tends to remain quite constant. Significant variation in the sum occurs predominately in the common mode. Thus, if the sum of the two emitter currents can be held constant, differential operation will not be significantly affected, since the changes in the two transistor currents are almost exactly opposite. Common-mode operation, however, causes both transistor currents to change in the same direction, thus causing a significant change in the total current through the emitter resistor. If the total is forced to remain at a constant figure, changes of both currents in the same direction will be suppressed, so that response to common-mode input signals will be reduced accordingly.

Some op amps utilize a simple but effective circuit placed in series with the shared emitter resistor to keep the shared emitter current quite constant. The circuit, called a **constant-current** circuit, is shown separately in Figure 1.6(a). Figure 1.6(b) shows the constant-current circuit as it is normally connected within the differential amplifier circuit. R_1, in the position of the differential amp's shared emitter resistor, provides current for the transistor's collector and also provides a small biasing current to R_2, which not only provides the transistor's base drive, but also forward-biases the diodes. R_3 carries the emitter current from the transistor.

Two series paths exist between the base terminal and the negative supply point: one path through the diodes represents the sum of their voltage drops, and the other path through the base-emitter junction and R_3 represents the sum of the voltage drop across the base-emitter junction plus the voltage across R_3. To satisfy Kirchhoff's voltage law, both paths must add up to the same total voltage. The voltage drop across D1 will very closely match the transistor's base-emitter voltage drop, since both are constructed from the same silicon wafer. Therefore, the voltage drop across R_3 will be required to match the

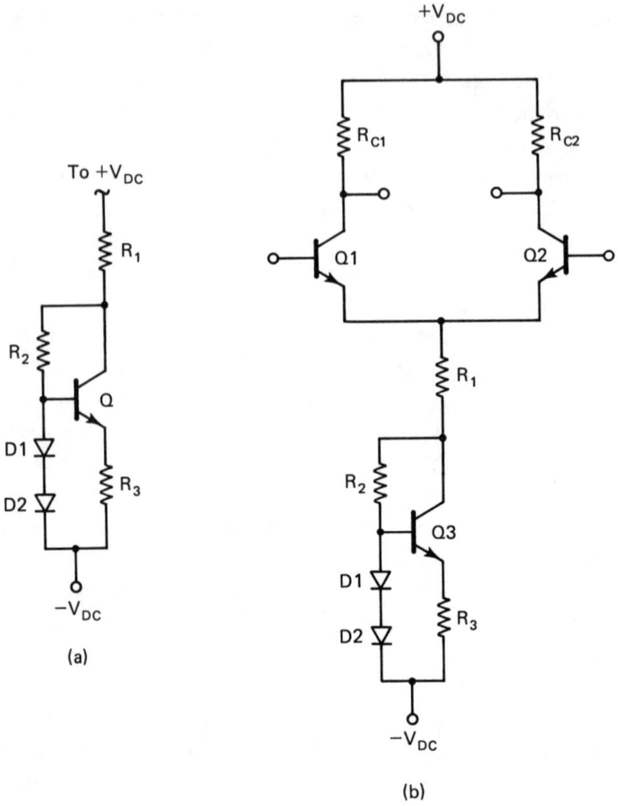

Figure 1.6

voltage drop across D2 in order to maintain proper bias conditions for the transistor.

Any change in emitter current will cause a similar change in the voltage across R_3. With an increase in current, the resistor's voltage rises, so the base-emitter voltage drop will decrease, which will decrease the transistor's base drive and collector current in compensation. In the case of a decrease in emitter current, the resistor's voltage falls, so the base-emitter voltage drop will increase, raising the transistor's base drive and collector current in compensation. The value of R_3 is critical, since it is responsible for setting the current level to be maintained. In some op amps, three or more diodes may be used in series, rather than just two. This will place a larger voltage across R_3, which allows a more precise control of the current level through it.

Op amps which utilize a constant-current circuit as part of the differential amplifier will generally provide very low common-mode gain figures. Since the op amp's differential gain, virtually unaffected by the presence of the constant-current circuit, will still be huge, op amps using the constant-current circuit within the first differential amp stage will also usually provide much larger *CMRR* specifications than other op amps.

1.6 INPUT AND OUTPUT IMPEDANCE OF THE DIFFERENTIAL AMPLIFIER

Since transistors are the active devices used in the construction of the differential amp, the impedance levels, logically, are also somewhat similar to other transistor-circuit impedance levels. Since both collector terminals function as the output connections, the output impedance will be roughly similar to the collector resistor value, just as in a common-emitter circuit. Since neither end of the output connects to ground, but rather to the opposite collector, the output impedance may actually be as large as double the value of the collector resistor. Since this value is generally of the order of many kilohms, the differential amp cannot deliver large output current levels, although it may develop a large output voltage amplitude with a sufficiently large resistance value connected in the load position.

The input impedance of the differential amp is also similar to that found in a common-emitter amplifier circuit containing an unbypassed emitter resistor. In such a circuit, the current gain of the transistor will cause the emitter resistor value to appear "beta" times larger when viewed from the base terminal, for an input impedance level of approximately "beta" times the emitter resistor value. In the differential amp, however, since an input at one base terminal will operate that transistor as a common emitter and the other transistor as a common base, the input impedance will usually be a bit smaller than that for a simple common-emitter circuit.

It is desirable for the differential amp to have as large an input impedance as possible. Since it is actually the first input stage of the op amp, the differential amp plays an almost exclusive role in determination of the op amp's input impedance. Several manufacturing methods have been employed to raise the input impedance of the differential amp to as large a figure as possible.

One simple way to maximize the impedance is to bias the differential amp circuit with current levels which are as small as possible. Differential amps will usually carry bias currents less than 100 μA and often less than 10 μA. Such tiny current levels result in large input impedances, of the order of hundreds of kilohms. This is a sufficiently large input impedance level for many op amp applications. The input impedance can be made even larger when necessary, through modification of the basic differential amp circuit.

In early op amps, each "side" of the differential amp often consisted not of a single transistor but rather of two transistors connected together in a **cascode** arrangement. A cascode amplifier consists of one transistor which operates as a common collector, driving a second transistor which operates as a common base. Together, both transistors provide an overall gain similar to that of a common emitter. Since the input is received by the common-collector section, the input impedance of this circuit can be significantly larger than that of the simple differential amplifier circuit with a single transistor on each side.

A more recent manufacturing technique involves constructing the differential amp with two Darlington pairs rather than two regular transistors. A **Darlington pair** consists of two bipolar junction transistors, with both collectors connected together, and the first transistor's emitter connected to the second transistor's base. Since the Darlington's external terminals consist of the first transistor's base, the second transistor's emitter, and the joined collectors, the Darlington's bias requirements and biasing circuits are quite similar to those of single transistors.

The current gain of a Darlington pair is effectively the product of the two separate transistor gain figures. This gives the device, as well as circuits in which it is used, immense overall gain characteristics. In addition, the input impedance figures of Darlington circuits may sometimes be dozens of times larger than those of simple two-transistor differential amp circuits, which is an additional effect resulting from the huge gain associated with the Darlington pair.

Another recent construction technique involves the use of two **field-effect transistors** in place of the usual bipolar-junction transistors, in the construction of the first differential amp circuit. A field-effect transistor, which requires no gate current under normal circumstances, provides a huge impedance level at its gate terminal, which also serves as the input connection in the common-source format. When the differential amp is constructed using two field-effect transistors, the circuit's input impedance will usually be several megohms or more in size.

Since the first input stage of the op amp is invariably a differential amplifier circuit, the input impedance of the differential amplifier becomes one of the most important external characteristics of the op amp in which it is used. Since a differential amp is never used to provide the final output of the op amp, but instead always provides the input signal for the next stage in the op amp, the output impedance level of the differential amp circuit is far less important for the op amp, since it has no discernible effect on the op amp's external impedance properties.

PROBLEMS

1.1. In a differential amplifier circuit, the transistor biasing circuitry most closely resembles which format?

(a) common base
(b) common emitter
(c) common collector

1.2. In a differential amplifier circuit, which transistor terminals are directly connected to a common resistor?

(a) The bases
(b) The collectors
(c) The emitters

1.3. In a differential amplifier circuit, the output signal connections involve which transistor terminals?

(a) The collectors
(b) The bases
(c) The emitters

1.4. In a differential amplifier circuit, when providing an input signal to only one of its input terminals, where will the other input terminal be connected?

(a) The output
(b) One DC supply
(c) Ground

1.5. In a differential amplifier circuit, when amplifying a single input signal, what phase relationship will exist between the signals at the two output terminals?

(a) A 180-degree difference
(b) A 90-degree difference
(c) In phase with each other
(d) No phase: only one signal is found

1.6. In a differential amplifier circuit, what phase relationship will always be found between two "common-mode" input signals?

(a) A 180-degree difference
(b) A 90-degree difference
(c) They are in phase with each other

1.7. In a differential amplifier circuit, what phase relationship will always be found between two "differential" input signals?

(a) A 180-degree difference
(b) A 90-degree difference
(c) They are in phase with each other

1.8. A "cascode" amplifier consists of two transistors which are connected in what first/second formats?

(a) Common collector to a common emitter
(b) Common emitter to a common base
(c) Common base to a common emitter
(d) Common collector to a common base
(e) Common emitter to a common collector

1.9. Which method of construction will normally provide the highest differential amplifier input impedance level?

(a) Using two Darlington pairs
(b) Using two field-effect transistors
(c) Using two junction transistors
(d) Using two cascode circuits

1.10. What is the purpose of including a constant-current circuit in a differential amplifier input stage?

(a) It increases differential gain
(b) It increases frequency response
(c) It decreases common-mode rejection ratio
(d) It decreases common-mode gain
(e) It decreases differential gain

1.11. An op amp is supplied with two 20-μV input signals. When the inputs are in phase, the output develops a 25-μV signal. When the inputs are 180 degrees out of phase, the output develops a 12-V signal. What are:

(a) A_{cm} (the common-mode gain figure)? ______

(b) A_{diff} (the differential gain figure)? ______

(c) *CMRR* (common-mode rejection ratio)? ______

1.12. An op amp is supplied with two 60-μV input signals. When the inputs are in phase, the output develops a 45-μV signal. When the inputs are 180 degrees out of phase, the output develops a 7.5-V signal. What are:

(a) A_{cm} (the common-mode gain figure)? ______

(b) A_{diff} (the differential gain figure)? ______

(c) *CMRR* (common mode rejection ratio)? ______

1.13. Specify and explain the phase relationship between a single input signal and the resulting signal found at the collector terminal of the:

(a) Transistor receiving the signal: ______

______

(b) Opposite transistor: ______

______

1.14. Specify and discuss the reasons why the addition of a constant-current circuit to a differential amplifier does not significantly affect response to differential input signals:

______

______

1.15. Specify and discuss the reasons why the addition of a constant-current circuit to a differential amplifier will significantly alter circuit response to common-mode input signals:

______

______

2

Operational Amplifier Characteristics

The op amp is in actuality a high-gain multistage amplifier circuit. Although most op amps actually contain more than three separate stages of amplification internally, three major functional blocks can commonly be identified in all op amps.

2.1 OP AMP CONFIGURATION

Figure 2.1 is a block diagram representing the functional contents of a typical op amp. The differential amplifier, as discussed in Chapter 1, always appears as the first stage of an op amp and is responsible for determination of the input impedance figure, as well as the availability of two inputs with different phase relationships. Many op amps actually contain multiple differential amp circuits, cascaded so that the output terminals of one stage are connected to both input terminals of the next stage. This type of connection serves to further enhance the amplification of differential inputs, while also further minimizing response to common-mode input signals. Since the differential amp provides a significant voltage gain, cascading more than one differential amp stage will help provide the huge overall voltage gain which is characteristic of the op amp.

The second basic portion of the op amp consists of voltage-amplification

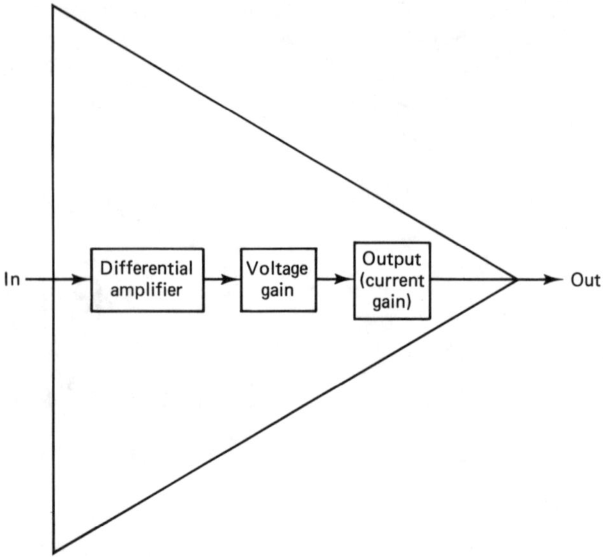

Figure 2.1

circuitry. This section may contain a differential amplifier stage, or it may instead be single transistor stages functioning in common-emitter or common-base modes, either of which can be used to supply voltage gain. This stage often develops the full output voltage amplitude, but since it has a fairly large output impedance level, only a small amount of output current can be safely supplied.

The third basic section of the op amp provides the current gain necessary in order to deliver large output currents to fairly small load impedance values, without loss of output voltage amplitude. Most op amps utilize a direct-coupled push-pull amplifier circuit as the final output section, allowing the output impedance level to be fairly small and output current levels to be fairly large. This output section of the op amp may consist of only the final output devices and their biasing circuitry, or it may be a more ornate circuit including driver transistor circuitry in addition to the output transistors themselves.

The standard schematic symbol for the op amp is shown in Figure 2.2.

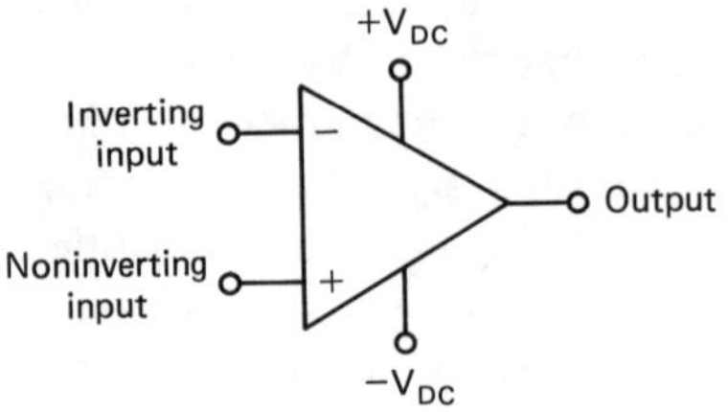

Figure 2.2

Always found together on the left side of the symbol, the two available input points will always be labelled. The "+" marking is used to designate the **noninverting** input, while the "−" marking designates the **inverting** input. Always found alone on the right side of the symbol is the *output* terminal. The two additional terminals placed above and below the center of the symbol represent the *DC voltage-supply connections*, with *positive above* and *negative below*.

2.2 IDEAL OP AMP CHARACTERISTICS

The op amp was first developed to approximate the conditions described in this section as closely as possible. Of course, like every other electrical component, the real op amp does not have "perfect" properties, but modern op amps show vast improvements over earlier versions in most areas. As manufacturing techniques become further refined, op amps continue to better approximate ideal conditions.

Gain. The ideal op amp has a gain figure which is infinitely large. Of course, no real op amp will ever have an infinite gain, but actual gain figures can be surprisingly large, of the order of one million or more for some types of op amps!

Common-Mode Rejection Ratio (*CMRR*). The ideal op amp will provide no response whatsoever to common-mode inputs, yet will provide an infinite gain for differential inputs. This leads to an ideal *CMRR* which is infinitely large, since *CMRR* equals differential gain divided by common-mode gain, as specified by equation (1.1).

Input Offset Voltage. The idea op amp will not require any **input offset voltage.** Input offset occurs in real op amps due to differences in the two transistors which make up the differential amplifier. If the two transistors have slightly different barrier potentials, then they will be biased at slightly different points by the shared emitter resistance. Since the collector currents will no longer be exactly equal, this will result in the presence of an offset voltage at the output, requiring the application of an input offset voltage to bring the output offset to zero. Of course, the ideal op amp would have identical transistors, in which case no offset voltage whatsoever would occur.

Input Impedance. The ideal op amp's input impedance figure should be infinitely large, so that no current would be required from the source of the input signal, insuring that the source will be able to maintain its full voltage amplitude. Since the input voltage variation will be maximum, maximum output voltage swing will also be insured by an infinite input impedance.

Output Impedance. The ideal op amp should be able to deliver any amount of current required by the load without any loss of voltage amplitude at the output. Therefore, its output impedance should be as small as possible: ideally, in fact, the ouptut impedance figure should be zero!

Slew Rate. The fastest possible rate of change that can occur in the output voltage is signified by the **slew rate.** The ideal op amp should be able to change its output instantaneously from one voltage level to another. This implies that the ideal slew rate is infinitely fast, involving no time delay whatsoever.

Frequency Response. The ideal op amp should be able to amplify any sort of input, from 0 Hz (DC) on up to any frequency in the AC realm. Thus, its frequency response should be infinitely wide, from DC up to infinite frequency figures.

Input Signal Limitations. The ideal op amp should be able to withstand any amplitude of input signal indefinitely, with no resulting damage occurring under any circumstances.

2.3 TYPICAL OP AMP CHARACTERISTICS

When we view the list of ideal characteristics in Section 2.2, it is easy to see that none of the ideal characteristics will ever truly be attained by an actual op amp. However, the typical op amp does provide characteristics which can allow it, when operating within its limitations, to provide circuit response characteristics which are surprisingly close to ideal predictions!

Gain. While the gain of an op amp can never be infinite, most op amps provide open-loop gain figures which are of the order of 100,000 when used alone, without the addition of any passive components or feedback paths. Among the wide variety of op amps produced, some types can provide open-loop gain of 1,000,000 or more!

CMRR. The common mode rejection ratio of an op amp could be infinite only if it provided no response at all to common-mode inputs. As discussed in Chapter 1, all op amps will show some minimal output response to common-mode inputs, but it will be sufficiently small to make a typical *CMRR* approximately 100,000 in size. This means that differential inputs will be amplified about 100,000 times as strongly as common-mode inputs of the same input amplitude. The larger the *CMRR* figure is, the smaller the amplification of common-mode inputs will be, as compared to the amplification of differential inputs.

On op amp data sheets, *CMRR* figures are sometimes provided in **decibel** form, rather than in resular decimal form. The number of decibels is determined by first finding the common log of the decimal *CMRR* figure, then multiplying the result by 20, according to this relationship:

$$CMRR \text{ in dB's} = 20 \cdot \log(CMRR \text{ in decimal form}) \tag{2.1}$$

EXAMPLE 2.1

(a) Convert a *CMRR* of 100,000 to decibels:

$$CMRR\ (\text{dB}) = 20 \cdot \log(100{,}000) = 20 \cdot 5 = 100 \text{ dB}$$

(b) Convert a *CMRR* of 2,000,000 to decibels:

$$CMRR\ (\text{db}) = 20 \cdot \log(2{,}000{,}000) = 20 \cdot 6.3 = 126 \text{ dB}$$

To convert a *CMRR* in decibel form back to the decimal version, first divide the number of decibels by 20, and then raise 10 to that power, according to this relationship:

$$CMRR \text{ in decimal form} = 10^{(\text{dB's}/20)} \tag{2.2}$$

EXAMPLE 2.2

(a) Convert a *CMRR* of 80 dB to decimal:

$$CMRR = 10^{(80/20)} = 10^4 = 10{,}000$$

(b) Convert a *CMRR* of 134 dB to decimal:

$$CMRR = 10^{(134/20)} = 10^{6.7} = 5{,}000{,}000$$

Input Offset Voltage. Since no two transistors are exactly alike, op amps will typically provide a tiny output voltage when both inputs are grounded. This tiny offset is predominately due to differences in barrier potential and gain between the transistors in the first differential amp stage. Amplification of this difference by the rest of the op amp results in the presence of the offset voltage at the output. Many op amps provide external terminals specifically for offset adjustment. Figure 2.3 shows the normal mode of connection of these terminals. They are connected one to each end of the fixed element in a potentiometer, while the wiper of the potentiometer is connected to the negative DC supply. Rotation of the wiper will cause variation in the biasing of the two transistors in the differential amp. This causes a similar variation in the output voltage, which is then adjusted as close to zero volts as possible. When it is necessary or desirable, the offset adjustment can also be used to provide desired small nonzero output DC offset levels.

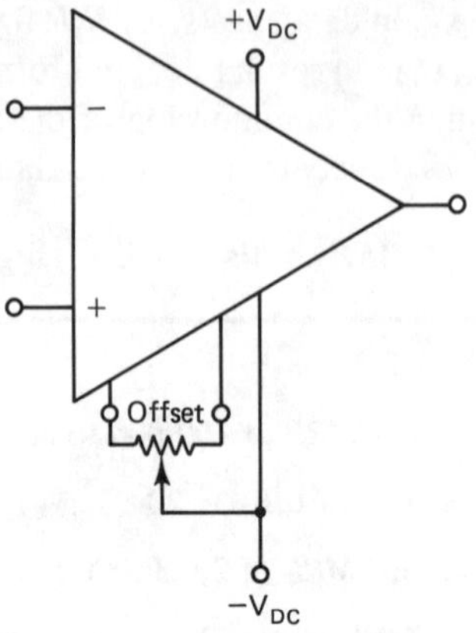

Figure 2.3

Input Impedance. While the ideal op amp should have an infinite input impedance, real op amps offer input impedance values which are usually above 100 kilohms. The use of Darlington pairs in the first differential amp stage, as discussed in Chapter 1, will normally provide input impedances greater than 1 megohm. Field-effect transistors, when used in the first differential amp stage, can provide input impedances of several megohms or more.

Output Impedance. No type of power-supply circuit, battery, or AC signal source can deliver an unlimited amount of current, a capability implied by the ideal output impedance figure of zero. In actuality, all sources have maximum current limitations, and they therefore have definite output impedance levels. The op amp is no different: any op amp has a maximum output current limit and thus a specific output impedance figure. With the variety of op amps available, output impedance values may be as large as several hundred ohms or as small as fifty ohms or less. In common types of op amps, a typical average value of output impedance would be near 100 ohms.

Slew Rate. Since transistors work through control of moving charges, which can neither start nor stop moving instantaneously, changes in current will always occur gradually, imposing a minimum time delay required to change voltage and current conditions. Op amps will always exhibit similar delays, since they contain transistors. Typical slew-rate figures range from a fraction of a volt to several volts per microsecond. Op amps with very fast slew rates will also normally offer better high-frequency response than those with slower slew rates.

Frequency Response. Every amplifying device functions only over a limited range of frequencies: no device yet known offers an unlimited bandwidth. Although some special op amps are designed to work at very high frequencies and may not function properly at lower frequencies, the majority of op amps are designed for low-frequency use, so they have only upper-frequency limitations. Low-frequency rolloff in amplifiers is usually caused by the presence of DC-blocking coupling devices, such as interstage coupling capacitors. The op amp, which is designed to operate with direct coupling at both input and output, is under these conditions capable of providing full gain to low-frequency signals, down to and including DC! Typical op amps offer frequency-response characteristics from zero hertz (DC) to upper limits of approximately 100 kilohertz or more.

Input Signal Limitations. Since its input stage will contain either field-effect transistors or bipolar-junction transistors, transistor breakdown may occur if an excessively large voltage is applied to either input terminal. In normal applications, input amplitudes should never exceed the DC supply values, remaining below the positive supply value and above the negative supply value at all times. This limit applies not only to AC signals for inputs but also to DC input levels as well.

2.4 DIRECT-CURRENT SUPPLIES FOR OP AMPS

The op amp is designed to be used with *dual* DC voltage supplies, which are normally of *equal voltage* but *opposite polarity*, with a common ground connection between them which is both the negative side of the positive supply and the positive side of the negative supply. Figure 2.2, the schematic symbol for the op amp, shows DC supply terminals in their standard positions. The op amp symbol will often be seen in schematic diagrams without the specific inclusion of DC supply terminals, but this does not mean that they are unused. The op amp, like any transistor, must have its DC power source for proper biasing or it will not function properly as an amplifier.

While the op amp can be used successfully with a single DC supply, the use of two equal positive and negative supplies allows the op amp to be direct-coupled at both the input and output connections. The op amp's internal biasing has been designed so that when dual supplies of equal size are used, the inputs and output will ride at ground level, halfway between the maximum and minimum supply values. Since this presents no DC levels which must be blocked, no coupling devices are necessary between successive op amp stages.

If we wish to operate the op amp from a single DC supply, we must observe certain limitations. The problem is a result of the conditions set by the internal biasing circuitry, which is arranged to place the input and output terminals at DC biasing levels which are halfway between the two supply-voltage values. With dual supplies, halfway between equal and opposite values falls at zero volts, or ground potential. However, with a single DC supply, the input and output terminals will now exhibit DC bias levels of approximately half the single supply-voltage value. Under these conditions, the DC levels present may require the use of coupling devices in order to achieve DC isolation between stages, which will introduce a low-frequency rolloff to the circuit's frequency response. Although single-supply op amp operation is not uncommon, most op amps operate in the dual-voltage-supply format for which they were originally designed when they are used as amplifiers.

PROBLEMS

2.1. What type of amplifier circuit will most commonly be used as the final output stage of an op amp?

(a) Differential amp
(b) Push-pull amp
(c) Voltage amp

2.2. When an op amp's inputs and output are direct-coupled, can the op amp amplify DC inputs?

(a) Yes, it can.
(b) No, it can't.

2.3. Which of the figures listed below best represents a typical op amp's open-loop gain figure?

(a) 1000
(b) 100,000
(c) 10,000,000
(d) 1,000,000,000

2.4. Which of the figures listed below best represents a typical op amp's input impedance level?

(a) 100 ohms
(b) 10 kilohms
(c) 1 megohm
(d) 100 megohms

2.5. Which of the figures listed below best represents a typical op amp's output impedance level?

(a) 1 ohm
(b) 10 ohms
(c) 100 ohms
(d) 1 kilohm

2.6. Which of the figures listed below best represents a typical op amp's slew-rate rating?

(a) 1 volt/microsecond
(b) 100 volts/microsecond
(c) 1 volt/millisecond
(d) 100 volts/millisecond

2.7. Which of the figures listed below best represents a typical op amp's frequency response?

(a) From 20 Hz to about 20 kHz
(b) From DC (0 Hz) to about 20 kHz
(c) From 20 Hz to about 200 kHz
(d) From DC (0 Hz) to about 200 kHz
(e) From DC (0 Hz) to infinite frequency values

2.8. We must make an adjustment in the offset voltage of an op amp. During this process, we should have a voltmeter connected between ground and which op amp terminal?

(a) The inverting input
(b) The noninverting input
(c) The output
(d) The positive DC voltage source
(e) None of these

2.9. If our op amp in problem 8 operates from dual positive and negative DC supplies, what voltage should appear on the meter when adjustment is complete?

(a) Zero volts
(b) Half the size of one DC supply
(c) The full DC supply value

2.10. If our op amp in problem 8 operates from a single positive DC supply, what voltage should appear on the meter when adjustment is complete?

(a) Zero volts
(b) Half the size of one DC supply
(c) The full DC supply value

2.11. For each of the following op amps, determine the *CMRR* figure, and state in both decimal and decibel forms:

(a) $A_{cm} = .8$ and $A_{diff} = 160{,}000$: $CMRR =$ ________ = ________ dB

(b) $A_{cm} = 1.5$ and $A_{diff} = 75{,}000$: $CMRR =$ ________ = ________ dB

(c) $A_{cm} = .6$ and $A_{diff} = 240{,}000$: $CMRR =$ ________ = ________ dB

(d) $A_{cm} = .3$ and $A_{diff} = 120{,}000$: $CMRR =$ ________ = ________ dB

(e) $A_{cm} = 1.2$ and $A_{diff} = 480{,}000$: $CMRR =$ ________ = ________ dB

(f) $A_{cm} = .9$ and $A_{diff} = 720{,}000$: $CMRR =$ ________ = ________ dB

2.12. Convert each *CMRR* from decibel form back to decimal form:

(a) 92 dB = ________________ **(b)** 106 dB = ________________

(c) 128 dB = ________________ **(d)** 114 dB = ________________

(e) 98 dB = ________________ **(f)** 132 dB = ________________

2.13. For each of the following *CMRR* figures stated in decibels, convert back to the decimal version of the *CMRR*, and then use it to find the missing gain figure:

(a) $CMRR = 128$ dB and $A_{\text{diff}} = 1{,}500{,}000$: $A_{\text{cm}} =$ ________________

(b) $CMRR = 114$ dB and $A_{\text{diff}} = 600{,}000$: $A_{\text{cm}} =$ ________________

(c) $CMRR = 108$ dB and $A_{\text{diff}} = 225{,}000$: $A_{\text{cm}} =$ ________________

(d) $CMRR = 92$ dB and $A_{\text{cm}} = .75$: $A_{\text{diff}} =$ ________________

(e) $CMRR = 118$ dB and $A_{\text{cm}} = .85$: $A_{\text{diff}} =$ ________________

(f) $CMRR = 106$ dB and $A_{\text{cm}} = .66$: $A_{\text{diff}} =$ ________________

3

Simple Op Amp Amplifier Circuits

The op amp is virtually never found alone in analog circuitry, where signals are normally amplified without introducing significant distortion. In the most common situation, at least two resistors will accompany the op amp. The reason is the huge gain figure which is characteristic of the op amp itself. No wideband amplifier can provide an output variation which exceeds its DC supply limits, and most amplifiers, including the op amp, will show evidence of signal clipping when their output swing exceeds 90% to 95% of the DC supply limits. In most analog circuits, particularly audio and radio, distortion should be reduced to as small an amount as possible.

A simple numerical example will illustrate the problem. Let us assume that we have an op amp, with a typical gain of 100,000, connected to two DC supplies of +12 V and −12 V. To minimize distortion, our output signal will be a maximum amplitude of 10 V peak. What input size will give us our maximum desired output? Since gain is output divided by input, then input size equals output size divided by gain. In this case,

$$V_{\text{in}} = \frac{V_{\text{out}}}{A} = \frac{10\text{ V}}{100{,}000} = 100\ \mu\text{V} \tag{3.1}$$

These calculations indicate that our maximum output will occur when our peak input size is only 100 μV! If our peak input exceeds 100 μV, our output will be driven into saturation, causing the output signal to be clipped.

Other problems can arise in conjunction with such huge gains. The tendency to drive the output into clipping is only one possibility. Another possibility is that of significant distortion, even in small output signals. We could also encounter severe limitation of upper-frequency response, which will be discussed in detail in Chapter 4. We might also discover that the circuit provides an output signal with no input signal present, a situation called **oscillation.** Any of these problems could occur in an amplifier with such a huge gain. In fact, the larger the gain figure is, the greater is the likelihood that they will occur!

3.1 FEEDBACK IN OP AMPS

The gain of any amplifier can be changed by the addition of **feedback** to the amplifier circuit. As shown in Figure 3.1, feedback occurs when a path exists to feed the output signal back to the input. Two types of feedback are possible.

In **positive feedback** a path is added to allow part of the output signal to return to the input in phase with the original input signal. The amplifier now receives a larger total signal at the input, leading to a larger output. Essentially, positive feedback will cause a larger output than would occur without the presence of the feedback path. Therefore, positive feedback *increases* the gain of the amplifier, leading to increased distortion, decreased frequency response, and increased tendency to oscillate — the potential problems discussed previously. Positive feedback is used in practice to construct oscillator circuits, but it is rarely used for any other type of application.

With **negative feedback** a path is added to allow part of the output signal to return to the input 180 degrees out of phase with the original input signal. The opposite phase will cause subtraction to occur between the original input signal and the feedback signal. The amplifier will now receive the smaller "difference" signal at its input, leading to a smaller output. Negative feedback will result in a smaller output than would occur without the feedback path. Therefore, negative feedback *decreases* the gain of the amplifier, which also serves to decrease distortion, minimize or eliminate the tendency to oscillate, and also widen frequency response! All three side effects, which were detrimental with positive feedback, have become advantageous with negative feedback, which for these reasons is commonly used in amplifier circuits.

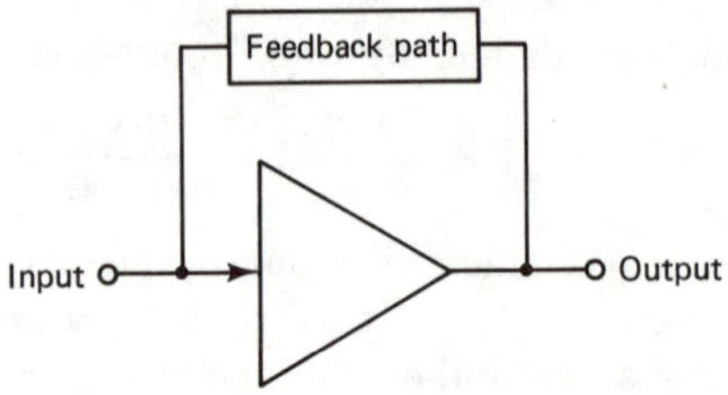

Figure 3.1

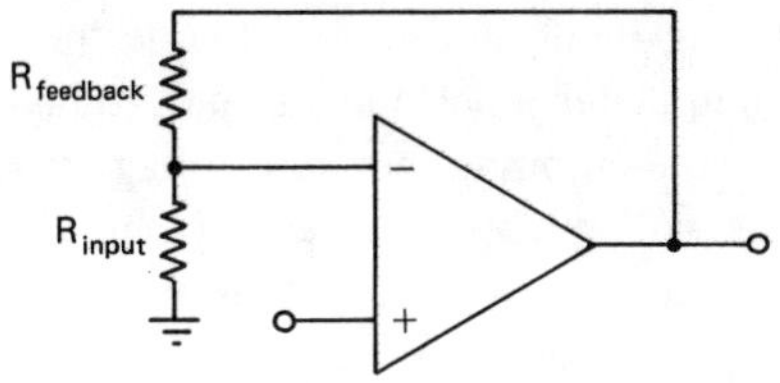

Figure 3.2

Negative feedback, used commonly to decrease the op amp's gain, must provide a portion of the output signal, through the feedback path, to the input of the amplifier, as well as leaving the original input source connected. In addition, it must be exactly 180 degrees out of phase (a reversal of polarity) with the original input. The simple way to accomplish this is to provide a resistive path from the output back to the inverting input, which by its nature is always 180 degrees out of phase with the output. Since only a portion of the output signal must be fed back, a simple voltage divider is used, as shown in Figure 3.2. Virtually all analog op amp circuits use this simple feedback path, regardless of the function of the circuit.

When this negative feedback path is present, the op amp will react in a very simple manner. Since both input points are directly involved with a forward-biased base-emitter junction, they need to be at the same, or at least very nearly the same, voltage. Through the feedback path, a portion of the output voltage is delivered to the inverting input. The output will, unless it reaches the DC supply limits, adjust itself to a voltage level necessary to keep both input terminal voltages *together*, with *no voltage difference* between them. This idea forms the basis for development of the mathematical relationships which describe op amp behavior.

3.2 UNITY-GAIN AMP OR BUFFER AMP

The simplest op amp circuit containing negative feedback is shown in Figure 3.3. The op amp's output terminal has been directly connected to the inverting input, forcing them to be always at the same voltage. The input signal is

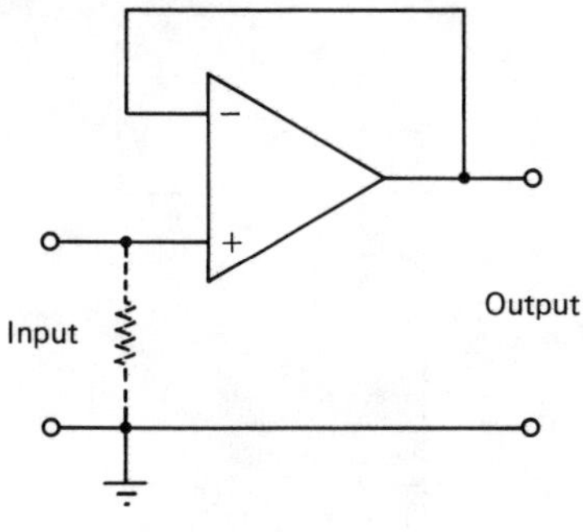

Figure 3.3

connected between the noninverting input and ground, while the output load, if used, connects between the output terminal and ground. When a signal is applied, a positive input will force the output to the same voltage level, resulting in both inputs being at the same voltage, as required. A negative input will cause the output to follow the input voltage again, thus keeping both inputs together. Simply stated, the output variation is identical in amplitude to the input variation, resulting in a gain figure of 1, or *unity*. In addition, the input and output signals will be in phase.

This circuit at first might seem rather useless, but the op amp is capable of providing a much larger current to an output load than existed at the input: while the voltage gain is 1, the circuit does provide a current-gain figure greater than one, similar in nature to a common collector amp, which also, with a voltage gain of 1, offers a large current-gain capability. The input and output impedances of this circuit are effectively those of the op amp itself, with an input impedance of 100 kilohms or more, and an output impedance of around 100 ohms. If a smaller input impedance is desired, a resistor may be added as shown in Figure 3.3 by connecting it between the noninverting input and ground. The input impedance will then be decreased to the parallel combination of the input impedance of the op amp itself with the ohm value of this additional resistor. If the resistor is significantly smaller than the op amp's input impedance, the input impedance of the circuit effectively becomes the resistor value alone.

3.3 THE NONINVERTING AMPLIFIER

The **noninverting** op amp circuit, shown in Figure 3.4, is similar to the unity-gain circuit, except the voltage-divider feedback path is used in its most common two-resistor form. The output will now deliver only a fraction of its voltage to the inverting input, requiring the output amplitude to be larger than the signal input in adjusting to keep both input terminal voltages together. Since the input

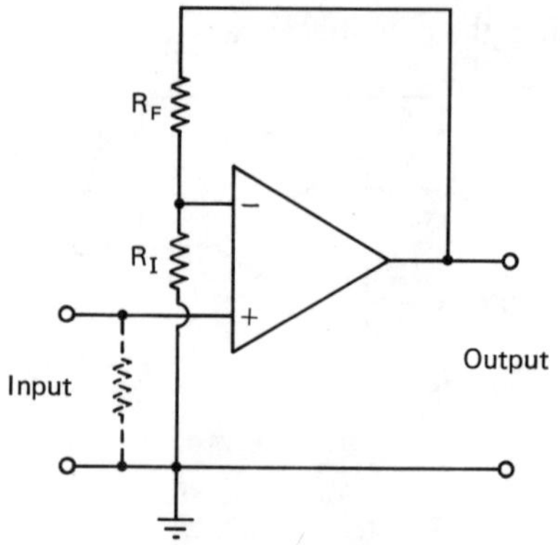

Figure 3.4

impedance of the op amp is extremely large, then the current through R_F will be approximately equal to the current through R_I. If we treat them as equal, then we can develop these relationships:

$$I_{R_F} = I_{R_I} \tag{3.2}$$

$$\frac{V_O - V_I}{R_F} = \frac{V_I}{R_I} \tag{3.3}$$

$$\frac{V_O - V_I}{V_I} = \frac{R_F}{R_I} \tag{3.4}$$

$$\frac{V_O}{V_I} - 1 = \frac{R_F}{R_I} \tag{3.5}$$

$$A_V = \frac{V_O}{V_I} = 1 + \frac{R_F}{R_I} \tag{3.6}$$

As stated by equation (3.6), the voltage gain of this noninverting amplifier is set solely by the two resistor values! When R_F is shorted, we have the minimum gain value of 1. When R_F contains resistance, the circuit gain will be greater than 1. By careful selection of resistor values, we can produce almost any gain figure that we might require, as long as it remains smaller than the op amp's own open-loop gain figure. Normally, these resistor values fall between 1 kilohm and 1 megohm.

EXAMPLE 3.1

Refer to Figure 3.4 for these questions:

(a) If R_F is 330 kilohms and R_I is 22 kilohms, what is the gain figure for the circuit?

$$A_V = 1 + \frac{330{,}000}{22{,}000} = 1 + 15 = 16$$

(b) If R_F is 110 kilohms and R_I is 22 kilohms, what is the gain figure for the circuit?

$$A_V = 1 + \frac{110{,}000}{22{,}000} = 1 + 5 = 6$$

(c) If R_F is 750 kilohms and R_I is 30 kilohms, what is the gain figure for the circuit?

$$A_V = 1 + \frac{750{,}000}{30{,}000} = 1 + 25 = 26$$

When the desired gain and one resistor have been selected, equation (3.6) can also be rearranged to allow selection of the second resistor. Solving for R_F yields:

$$R_F = R_I \cdot (A_V - 1) \tag{3.7}$$

This equation says the proper value for R_F must be the value of R_I multiplied by the desired gain figure minus 1. Solving equation (3.6) for R_I yields:

$$R_I = \frac{R_F}{A_V - 1} \tag{3.8}$$

This equation says the proper value for R_I must be the value of R_F divided by the desired gain figure minus 1.

EXAMPLE 3.2

Refer to Figure 3.4 for these questions:

(a) We want a gain of 9 with a value for R_I of 15 kilohms. What must the value of R_F be?

$$R_F = 15\ \text{k}\Omega \cdot (9 - 1) = 15\ \text{k}\Omega \cdot 8 = 120\ \text{k}\Omega$$

(b) We want a gain of 6 with a value for R_F of 100 kΩ. What must the value of R_I be?

$$R_I = \frac{100\ \text{k}\Omega}{6 - 1} = \frac{100\ \text{k}\Omega}{5} = 20\ \text{k}\Omega$$

(c) We want a gain of 7.2 with a value for R_I of 100 kilohms. What must the value of R_F be?

$$R_F = 100\ \text{k}\Omega \cdot (7.2 - 1) = 100\ \text{k}\Omega \cdot 6.2 = 620\ \text{k}\Omega$$

(d) We want a gain of 25 with a value for R_F of 1.2 megohms. What must the value of R_I be?

$$R_I = \frac{1.2\ \text{M}\Omega}{25 - 1} = \frac{1.2\ \text{M}\Omega}{24} = 50\ \text{k}\Omega$$

As the amplifier's name indicates, input and output will be in phase with each other, or not inverted. Input and output impedance levels are still approximately the op amp's own impedance values, although if a smaller input impedance is required, a resistor can be added between the noninverting input and ground, causing the input impedance to decrease to approximately that resistance value, in the same way as was shown in unity-gain amp.

3.4 THE INVERTING AMPLIFIER

The **inverting** op amp circuit, shown in Figure 3.5, uses the same two feedback elements as the noninverting amp. There is, however, one major difference in the connection of the input signal source between ground and the end of R_I that was grounded in the noninverting amp, so that now the noninverting input, being unused, is grounded instead. This reversal of input and ground positions results in a different type of response from the op amp.

We now have the noninverting input grounded. The op amp's output, in keeping the inputs together, will always try to make the inverting input stay at ground level as well. This has two interesting effects. First, since the inverting input will always be held at ground potential, the input impedance is now effectively the value of R_I, rather than that of the op amp. Second, since input and output are at opposite ends of the resistor string, in which the middle connection must remain at ground potential, the two ends of the resistor string must therefore be opposite in polarity at all times, as the name "inverting" implies. As before, both resistor currents will be equal, from which we can develop these relationships:

$$I_{R_F} = I_{R_I} \tag{3.9}$$

$$\frac{V_O - 0}{R_F} = \frac{0 - V_I}{R_I} \tag{3.10}$$

$$\frac{V_O}{R_F} = -\frac{V_I}{R_I} \tag{3.11}$$

$$A_V = \frac{V_O}{V_I} = -\frac{R_F}{R_I} \tag{3.12}$$

As stated by equation (3.12), the voltage gain of this inverting amp is also determined solely by the two resistor values! The negative sign is a mathemat-

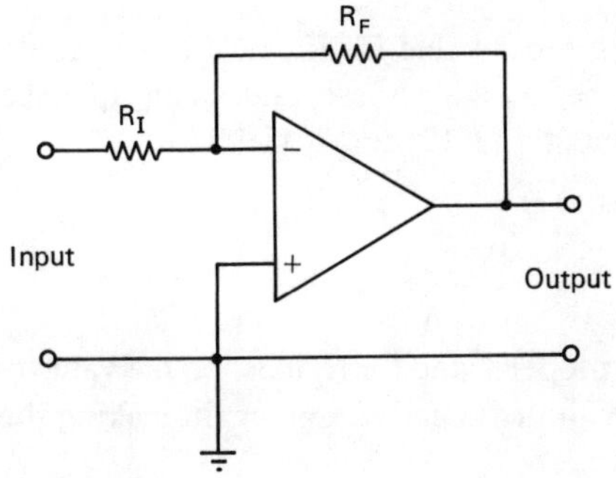

Figure 3.5

ical indication that input and output will be 180 degrees out of phase, rather than in phase.

EXAMPLE 3.3

Refer to Figure 3.5 for these questions:

(a) If R_F is 180 kilohms and R_I is 15 kilohms, what is the gain figure for the circuit?

$$A_V = -\frac{180,000}{15,000} = -12$$

(b) If R_F is 300 kilohms and R_I is 12 kilohms, what is the gain figure for the circuit?

$$A_V = -\frac{300,000}{12,000} = -25$$

(c) If R_F is 10 kilohms and R_I is 20 kilohms, what is the gain figure for the circuit?

$$A_V = -\frac{10,000}{20,000} = -.5$$

As shown by Example 3.3(c), the inverting format, unlike the noninverting one, is capable of providing gain figures smaller than 1, in which case the output will be smaller in amplitude than the input. This is accomplished by making the value of R_F smaller than that of R_I.

When the desired gain figure and one resistor have already been selected, equation (3.12) can be rearranged to allow selection of the other resistor value. Solving for R_F yields:

$$R_F = R_I \cdot -A_V \tag{3.13}$$

This equation states that the proper value for R_F must be the value of R_I times the desired gain figure. The negative sign associated with A_V makes it positive for this purpose. Solving equation (3.12) for R_I yields:

$$R_I = \frac{R_F}{-A_V} \tag{3.14}$$

This equation states that the proper value for R_I must be the value of R_F divided by the desired gain figure, with the negative sign again making the gain figure positive for this purpose.

EXAMPLE 3.4

Refer to Figure 3.5 for these questions:

(a) We want a gain of -15 with a value for R_F of 300 kilohms. What must the value of R_I be?

$$R_I = \frac{300\text{ k}\Omega}{-(-15)} = \frac{300\text{ k}\Omega}{15} = 20\text{ k}\Omega$$

(b) We want a gain of -5.5 with a value for R_I of 22 kilohms. What must the value of R_F be?

$$R_F = 22\text{ k}\Omega \cdot -(-5.5) = 22\text{ k}\Omega \cdot 5.5 = 110\text{ k}\Omega$$

(c) We want a gain of -12 with a value for R_F of 1.8 megohms. What must the value of R_I be?

$$R_I = \frac{1.8\text{ M}\Omega}{-(-12)} = \frac{1.8\text{ M}\Omega}{12} = 150\text{ k}\Omega$$

(d) We want a gain of -25 with a value for R_I of 22 kilohms. What must the value of R_F be?

$$R_F = 22\text{ k}\Omega \cdot -(-25) = 22\text{ k}\Omega \cdot 25 = 550\text{ k}\Omega$$

As previously stated, this format's input impedance is the resistance value of R_I and will usually be much smaller than that of the noninverting amp. This format's output impedance, because R_F provides an extra path to ground, will be the parallel combination of R_F and the output impedance value for the op amp itself.

3.5 OUTPUT SIGNAL AMPLITUDE LIMITATIONS

The use of negative feedback with the op amp serves to decrease its gain, decrease distortion, minimize oscillation tendencies, and expand the frequency response. When R_F and R_I are used, they determine the gain of the circuit. Any amplifier will attempt to drive its output as far as the input and gain prescribe. However, the output can never go beyond the DC values and will normally limit at one to two volts below maximum and above minimum. When the output is driven to either extreme, a part of the waveform will be "clipped," or simply not amplified, which seriously distorts the output wave. Under these conditions, the gain figure becomes meaningless as well: the gain figure can be valid only when the output is kept below amplitudes causing clipping. Op amp inputs should always be kept within sufficiently small amplitudes to provide only undistorted output signals, below clipping amplitudes, unless a clipped output is desirable.

3.6 TROUBLESHOOTING

When problems occur in an op amp circuit, one of two symptoms will usually be seen: no output signal appears, or a very distorted output signal appears. In the case of no output, make sure that your DC supplies are active, and make sure that your input signal source is functional. Check bias levels on the op amp's inputs and output: all three should be at ground with dual supplies, or at half the supply voltage with a single supply. When DC voltages and input signal are correct but no output appears, the op amp will usually be faulty, unless a defective or missing connection exists.

When an output does appear but it is seriously distorted, first try decreasing the input size, which should decrease the output size as well. This will resolve the problem if it was simply too much input signal amplitude. However, if one alternation seems to be missing, check DC supplies and input signal: when one of two source voltages is missing, one side of the output will also be missing, and if the input is distorted, the output can be no better. If the output terminal seems to show a DC voltage level, attempt to readjust the offset potentiometer for an output DC level of zero. If it can't be adjusted or if it doesn't reach zero, the op amp has probably been damaged internally.

PROBLEMS

3.1. When positive feedback is used in an amplifier circuit, which of the following effects will be expected too?
(a) Increased gain
(b) Decreased frequency response
(c) Increased distortion
(d) Tendency to oscillate
(e) All of the above
(f) None of the above

3.2. When negative feedback is used in an amplifier circuit, which of the following effects will be expected too?
(a) Decreased gain
(b) Increased frequency response
(c) Decreased distortion
(d) Will not oscillate
(e) All of the above
(f) None of the above

3.3. In the unity-gain op amp circuit, what is the phase relationship between the input and output signals?
(a) In phase
(b) A 90-degree shift
(c) A 180-degree shift

3.4. What type(s) of gain can normally be expected from the unity-gain op amp circuit?
(a) Resistance gain
(b) Current gain
(c) Power gain
(d) (a) and (b) only
(e) (a) and (c) only
(f) (b) and (c) only

3.5. Which of the op amp formats listed below will normally provide the smallest input impedance?
(a) Inverting
(b) Noninverting
(c) Unity gain

3.6. Which of the op amp formats listed below provides an input impedance which is approximately equal to the value of R_I in the feedback path?
(a) The inverting amplifier format
(b) The noninverting amplifier format
(c) The unity-gain amplifier format
(d) None of the above

3.7. Referring to the circuit in Figure 3.4, find the gain figure corresponding to each pair of resistor values listed:

	R_F	R_I	A_V		R_F	R_I	A_V
(a)	270 kΩ	10 kΩ	______	**(b)**	3.3 MΩ	200 kΩ	______
(c)	390 kΩ	12 kΩ	______	**(d)**	10 kΩ	22 kΩ	______
(e)	82 kΩ	51 kΩ	______	**(f)**	12 MΩ	300 kΩ	______

3.8. Referring to the circuit in Figure 3.4, find the missing resistor value needed to accompany each gain and resistance given:

	R_F	R_I	A_V		R_F	R_I	A_V
(a)	120 kΩ	______	25	**(b)**	______	18 kΩ	10
(c)	______	7.5 kΩ	5	**(d)**	950 kΩ	______	20
(e)	680 kΩ	______	18	**(f)**	______	10 kΩ	4.3

3.9. Referring to the circuit in Figure 3.5, find the gain figure for each pair of resistor values listed:

	R_F	R_I	A_V		R_F	R_I	A_V
(a)	160 kΩ	10 kΩ	______	**(b)**	750 kΩ	30 kΩ	______
(c)	5.6 MΩ	180 kΩ	______	**(d)**	680 kΩ	68 kΩ	______
(e)	12 kΩ	30 kΩ	______	**(f)**	56 kΩ	16 kΩ	______

3.10. Referring to the circuit in Figure 3.5, find the missing resistor value needed to accompany each gain and resistance given:

	R_F	R_I	A_V		R_F	R_I	A_V
(a)	270 kΩ	______	−9	**(b)**	______	20 kΩ	−5.5
(c)	______	13 kΩ	−30	**(d)**	680 kΩ	______	−17
(e)	3.3 MΩ	______	−15	**(f)**	______	15 kΩ	−8

4 Other Op Amp Amplifier Circuits

We have seen in the last chapter that both the inverting and the noninverting op amp circuits have gain figures which are determined solely by the values of the two external feedback resistors R_F and R_I. For the noninverting amp, the gain figure was $(R_F/R_I) + 1$, so even if R_F is a smaller value than R_I, the gain is still larger than 1. When R_F is reduced to zero ohms, or a short, we have the unity-gain circuit. In the inverting amp, the gain figure was $-R_F/R_I$, so if we want a gain larger than 1, R_F must be larger than R_I. If R_F is smaller than R_I, the gain will be less than 1, providing an output which is smaller than the input. If R_F is reduced to a short, the gain becomes zero, providing no output signals even when inputs are provided. The negative sign in the gain formula signifies the out-of-phase relationship between input and output signals. If R_F is large compared to R_I, either format will provide a large gain.

4.1 GAIN AND FREQUENCY RESPONSE

Figure 4.1 shows two **multistage** op amp circuits. Circuit (a) consists of two cascaded stages, each having a gain figure of 100, while circuit (b) has four cascaded stages, each providing a gain figure of 10. Overall system gain in a multistage amp is found by multiplying the individual stage gains together, so

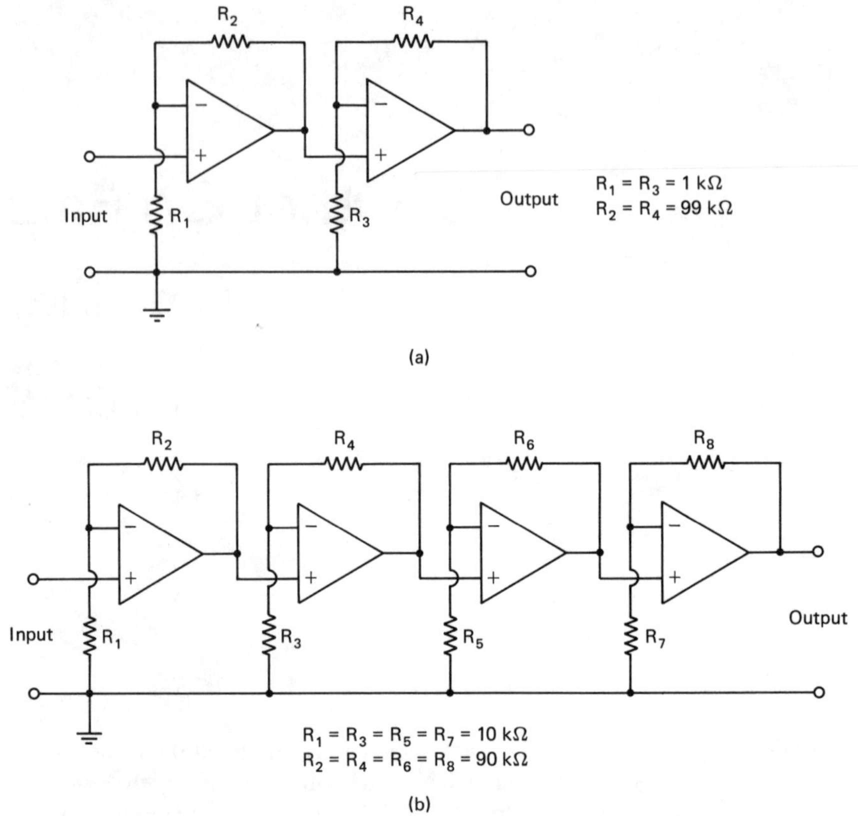

Figure 4.1

both amplifiers will provide the same overall system gain figure of 10,000. Clearly, circuit (a) is much simpler than circuit (b), since it contains only half as many components. Circuit (a) will thus be less costly to construct. Does this make it preferable to circuit (b)? If we constructed both circuits, would they provide identical features?

If we actually did build both circuits and measured such features as gain, distortion, and frequency response, we would find both system gains to be the same. However, the two systems would not provide identical distortion and frequency-response properties. Circuit (a) would exhibit more distortion and a much narrower frequency response than circuit (b). It would then seem that circuit (b) may be preferable to circuit (a), since it gives the same gain, over a wider frequency range and with less distortion! But why do these differences in distortion and bandwidth occur?

We have previously discussed the fact that a higher op amp gain will also normally be accompanied by a larger distortion level. Circuit (b), having lower gain per stage, will also exhibit less distortion per stage. The fact that we have twice as many stages in circuit (b) has very little significance: circuit (a) will exhibit more distortion in each stage, and therefore in the entire system, as a result of its larger individual stage gain. Clearly, huge gains with small levels of distortion can be realized by cascading enough low-gain stages. But why is frequency response also affected?

Figure 4.2 shows the frequency response of a typical op amp, operating *open-loop* without any feedback. At low frequencies, its gain is about 100,000. However, above 10 Hz, the upper rolloff frequency, the gain decreases as we go up in frequency, until at 1 MHz the gain has fallen to 1! Above 1 MHz the gain of the op amp is smaller than 1.

Figure 4.3 shows the same frequency response, with two additional lines depicted. These lines represent the *feedback-limited* gains of 100 (the upper line) and 10 (the lower line). Note that for each lower gain figure, the upper rolloff frequency is correspondingly higher! With gain at 100, upper rolloff is at 10 kHz, while for a gain of 10, upper rolloff is at 100 kHz. When negative feedback reduces the gain to a smaller figure, the gain of the op amp itself is larger than the imposed gain figure over a much wider range of frequencies.

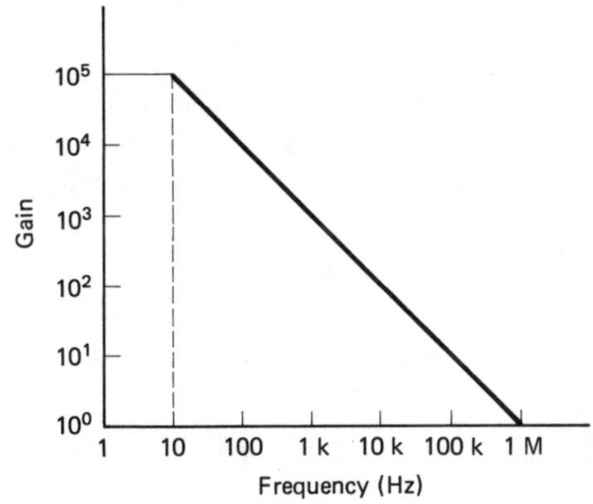

Figure 4.2

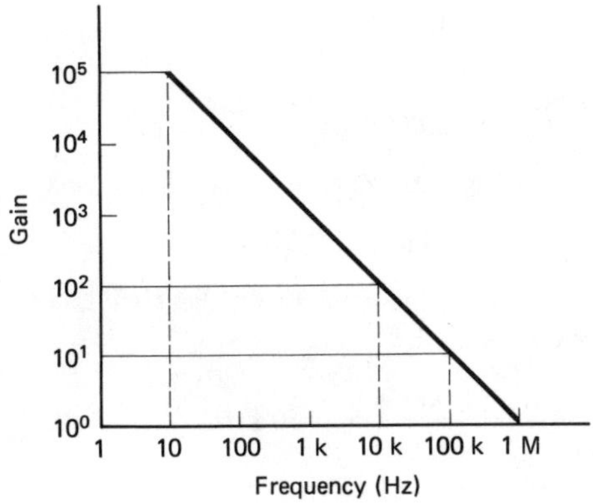

Figure 4.3

Since gain will be limited to the imposed value, this gain will result for all frequencies at which the op amp's gain figure is equal or higher. The lower we make the imposed gain, the higher the frequency will be at which the op amp's gain equals imposed gain. Therefore, when gain is made smaller, frequency response becomes correspondingly larger.

An op amp specification called **gain-bandwidth product,** or ***GBW,*** is specifically related to both gain and frequency response, or bandwidth, as its name suggests. *GBW* is defined as the result of amplifier gain times bandwidth:

$$GBW = A_V \times BW \tag{4.1}$$

For any op amp, *GBW* is a constant figure, so if it and the gain are known, the resulting bandwidth can be quickly calculated:

$$BW = \frac{GBW}{A_V} \tag{4.2}$$

When the *GBW* is a known figure, the appropriate gain figure which will provide a desired bandwidth can be quickly determined using equation (4.3):

$$A_V = \frac{GBW}{BW} \tag{4.3}$$

The *GBW* value can also be interpreted as the frequency at which the op amp's gain falls to 1. In our previous example, *GBW* is a constant 1,000,000, with open-loop gain falling to 1 at 1 MHz. With negative feedback added to the circuit, a gain of 100 times a bandwidth of 10 kHz still gives a *GBW* of 1,000,000, as a gain of 10 times a bandwidth of 100 kHz.

EXAMPLE 4.1

Let us assume that we have an op amp with a *GBW* of 2,000,000 and an open-loop gain figure of 50,000.

(a) What is the open-loop bandwidth?

$$BW = \frac{2{,}000{,}000}{50{,}000} = 40 \text{ Hz}$$

(b) At what frequency will the op amp's open-loop gain figure be a value of 1?

Gain falls to 1 at 2 MHz

(c) If negative feedback imposes a gain of 40 on the circuit, what will the bandwidth or frequency response be?

$$BW = \frac{2{,}000{,}000}{40} = 50 \text{ kHz}$$

(d) If we desire a bandwidth of 100 kHz, what would the gain figure of the circuit need to be?

$$A_V = \frac{2{,}000{,}000}{100{,}000} = 20$$

As indicated by these examples, any change in gain is always accompanied by an opposite change in bandwidth.

4.2 MULTIPLE-INPUT OP AMP AMPLIFIERS

By using an op amp in the differential mode, both input terminals are used concurrently to allow the simultaneous inputting of two signals. However, another op amp format allows multiple inputs to be connected and amplified, with no maximum limit on the number of inputs that can be included. This circuit uses neither differential nor common-mode inputs: each input signal is connected to the inverting input through a separate series resistor R_I, uniquely functioning for that signal input only.

This type of circuit, shown in Figure 4.4, is called a **mixer** or **summing amp** when the R_I's are all the same value, with the following formula describing its behavior:

$$V_O = -\frac{R_F}{R_I} \cdot (V_1 + V_2 + V_3 + \cdots) \tag{4.4}$$

With all R_I values the same, every input point receives the same gain as every other input, so that all signals are amplified equally, and the output is the direct sum of all input signals.

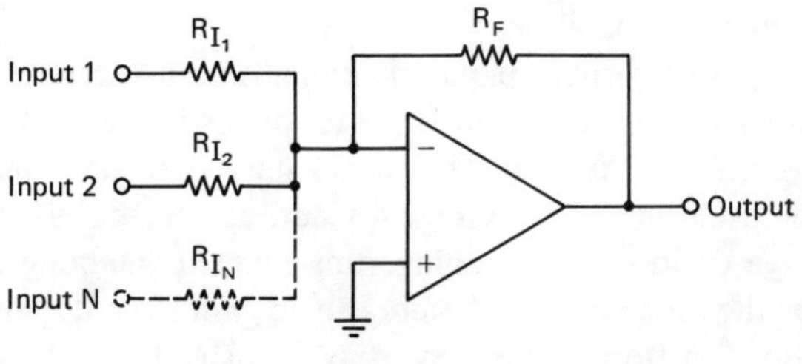

Figure 4.4

EXAMPLE 4.2

If R_F in Figure 4.4 is 150 kilohms and each R_I is 10 kilohms, every input will receive the same gain of -15 and provide an input impedance level of 10 kilohms.

Note that with only one input, equation (4.4) would simplify to exactly match the form of the equation governing the simple inverting amplifier. Based on the inverting format, mixers also exhibit other inverting properties, such as a phase reversal between input and output. Input impedance levels, too, are determined by the R_I value used at each input point.

The circuit is instead called a **weighted summing amp** when the R_I's are different values. In this case each input point will receive its own gain, determined by the R_I value at that particular input. The following formula describes this circuit's behavior:

$$V_O = -\frac{R_F}{R_1} \cdot V_1 - \frac{R_F}{R_2} \cdot V_2 - \cdots \tag{4.5}$$

EXAMPLE 4.3

If R_F in Figure 4.4 is 100 kilohms, R_{I_1} is 10 kilohms, and R_{I_2} is 20 kilohms, signals at input 1 will receive a gain of -10 with an input impedance of 10 kilohms, while signals at input 2 will receive a gain of only -5 but with an input impedance of 20 kilohms.

Note that the weighted mixer will provide different impedances at different inputs.

If a potentiometer, using the wiper terminal and one end of the fixed element, is connected to function as R_F instead of a fixed resistor, the gain and the resultant output amplitude of the op amp become adjustable through variation of the potentiometer. Capable of varying the inverting gain between zero and a maximum of $-R_F/R_I$, the output amplitude can therefore be adjusted between zero and maximum amplitude through movement of the potentiometer. With its ability to mix multiple input signals at a single output, the mixer is often used for audio applications where output amplitude adjustment is desired in addition to mixing capabilities.

Another op amp format, popularly called the **difference amp,** is shown in Figure 4.5. This circuit uses both input terminals in the differential mode, as the name suggests, and it therefore allows only two input points. Note the presence of the usual feedback voltage divider, and in addition the presence of a second voltage divider at the noninverting input, containing the same resistance values as the other voltage divider. Since gain from the inverting input is $-R_F/R_I$, while gain from the noninverting input is $1 + (R_F/R_I)$, the noninverting input's voltage divider will present $R_F/(R_F + R_I)$ times its signal source

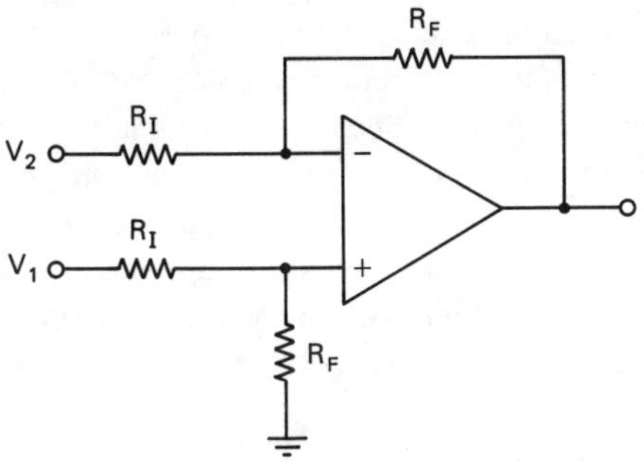

Figure 4.5

to the actual op amp input. Comparing the output with that input source directly, the effective gain will be:

$$A_V = \left(1 + \frac{R_F}{R_I}\right) \cdot \frac{R_F}{R_F + R_I} \quad (4.6)$$

$$A_V = \left(\frac{R_I + R_F}{R_I}\right) \cdot \frac{R_F}{R_F + R_I} \quad (4.7)$$

$$A_V = \frac{R_F}{R_I} \quad (4.8)$$

This means that both inputs will receive the same overall gain, so that the output will be a direct result of the difference between the two inputs. In the construction of this circuit, both resistors labeled R_F are the same resistance, and both labeled R_I are also the same, insuring that both inputs receive the same overall gain. The governing formula for this circuit is:

$$V_O = \frac{R_F}{R_I} \cdot (V_1 - V_2) \quad (4.9)$$

EXAMPLE 4.4

In Figure 4.5, assume that both R_F's are 30 kilohms and both R_I's are 10 kilohms.

(a) $V_1 = +1.5$ V and $V_2 = +.5$ V: What is V_O?

$$V_O = \frac{30\text{ k}\Omega}{10\text{ k}\Omega} \cdot (1.5 - .5) = 3 \cdot 1 = +3\text{ V}$$

(b) $V_1 = +.4$ V and $V_2 = -1.2$ V: What is V_O?

$$V_O = \frac{30\text{ k}\Omega}{10\text{ k}\Omega} \cdot [.4 - (-1.2)] = 3 \cdot 1.6 = +4.8\text{ V}$$

(c) $V_1 = -3$ V and $V_2 = -1.4$ V: What is V_O?

$$V_O = \frac{30\text{ k}\Omega}{10\text{ k}\Omega} \cdot [-3 - (-1.4)] = 3 \cdot -1.6 = -4.8\text{ V}$$

The difference amp, since it actually utilizes both inputs, will provide a gain which depends not only on the feedback resistor values, but also on the phase relationship between both input signals. When the two signals are out of phase, a large output will be developed by differential mode action. When the two input signals are in phase, however, common-mode action results in little or no output signal. The difference amp, actually functioning in the differential mode, responds only to differences between its two inputs. As equation (4.8) implies, identical inputs would make V_1 and V_2 equal, so that $V_1 - V_2$ would be zero.

4.3 TROUBLESHOOTING OP AMP AMPLIFIERS

If a problem develops with a multistage amplifier, similar procedures are used regardless of whether the output is missing or distorted. First, check DC voltage sources. Next, verify that an input is being applied to the first stage. Now, move halfway to the output (in an eight-stage amp, go to the fourth stage) and check for signal presence and shape. If it looks good, go halfway toward the output and check the waveform again. If it looks bad, go halfway back to the input and check the waveform. Continue moving halfway toward the input or the output until the defective stage is located.

If a problem arises with a summing or difference amp, your procedure depends on what type of output is present. If the output is clipped, decrease input sizes. If no output or a distorted output is present, check DC supplies and chip bias levels. Readjust offset if needed. If other conditions are correct, the op amp is probably defective. If some inputs respond properly but others don't, check input resistors and solder connections. If possible, use each input one at a time to locate the difficulty.

PROBLEMS

4.1. Find the total system gain figure for each multistage amplifier circuit shown in Figure 4.6:

(a) $A_V =$ ______ **(d)** $A_V =$ ______
(b) $A_V =$ ______ **(e)** $A_V =$ ______
(c) $A_V =$ ______ **(f)** $A_V =$ ______

4.2. We are working with an operational amplifier which has a gain-bandwidth product of 2,400,000.

(a) If open-loop gain is 80,000, bandwidth = ______

(b) For a bandwidth of 20 kHz, gain = ______

(c) If the gain is reduced to 15, bandwidth = ______

(d) For a bandwidth of 100 kHz, gain = ______

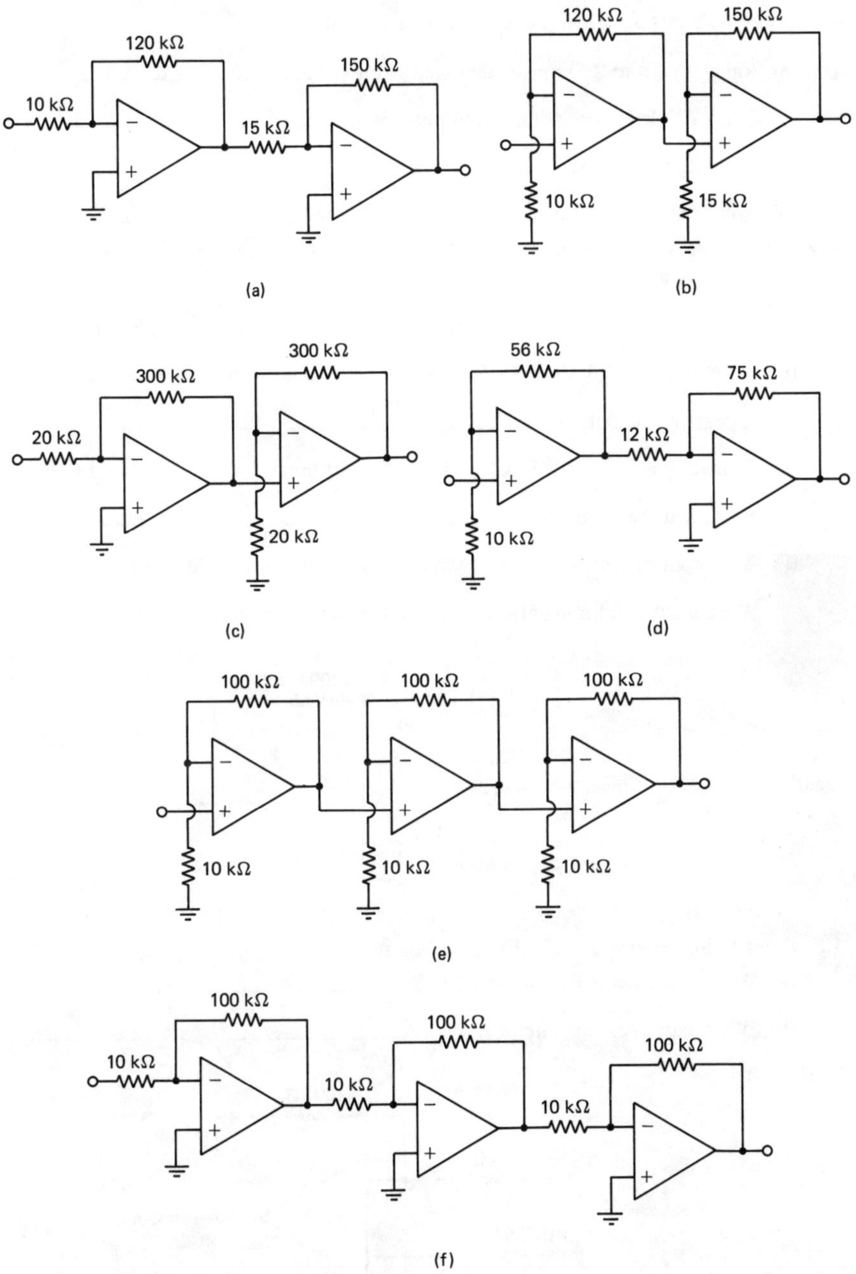
120 kΩ
10 kΩ
15 kΩ
150 kΩ
(a)
120 kΩ
150 kΩ
10 kΩ
15 kΩ
(b)
300 kΩ
20 kΩ
300 kΩ
20 kΩ
(c)
56 kΩ
12 kΩ
75 kΩ
10 kΩ
(d)
100 kΩ
100 kΩ
100 kΩ
10 kΩ
10 kΩ
10 kΩ
(e)
100 kΩ
10 kΩ
10 kΩ
100 kΩ
10 kΩ
100 kΩ
(f)

Figure 4.6

4.3. An op amp has a 35-kHz bandwidth with gain of 40.

(a) Raising gain to 200 makes the bandwidth = ________

(b) For a 200-kHz bandwidth, gain must = ________

(c) Gain decreases to 1 when the frequency = ________

(d) Lowering gain to 2.5 makes the bandwidth = ________

4.4 Refer to the op amp circuit shown in Figure 4.7. What gain figure will be received by the signal at:

(a) Input 1?___ **(b)** Input 2?___ **(c)** Input 3?___

(d) If input 1 = +.5 V, input 2 = −1 V, and input 3 = −.3 V, what voltage appears at the output?________

(e) If input 1 = −1.8 V, input 2 = +3 V, and input 3 = +.9 V, what voltage appears at the output?________

(f) An output of zero volts results when input 1 = +.7 V and input 3 = −2 V. What input level must also be supplied to input 2 to make this true?________

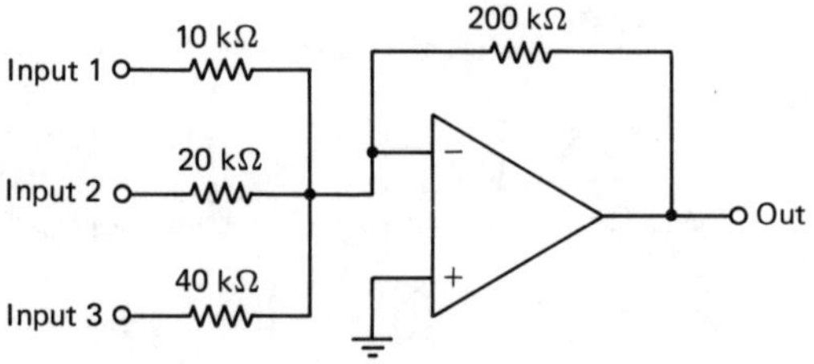

Figure 4.7

4.5. Refer to the op amp circuit shown in Figure 4.8.

(a) What is the name of this circuit?________

(b) What gain does this circuit provide?________

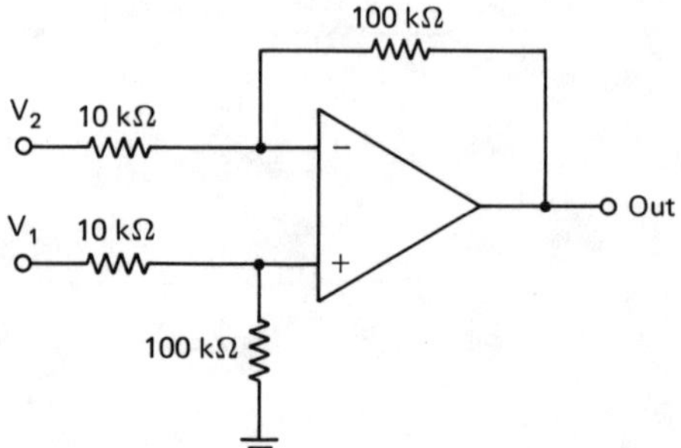

Figure 4.8

(c) If $V_1 = -2$ V and $V_2 = -3.5$ V, output voltage = ________________

(d) If $V_1 = 1.3$ V and $V_2 = 2.1$ V, output voltage = ________________

(e) If $V_1 = +.2$ V and $V_2 = -.2$ V, output voltage = ________________

(f) With AC signals to both inputs, which situation will deliver the largest output signal amplitude?________________

(a) Inputs in phase
(b) A 90-degree phase difference
(c) A 180-degree phase difference

5

Passive and Active Filters

Both passive and active filters are commonly found in *low pass, high-pass, band-pass,* and *band-reject* or *band-stop* types. There is, however, one major difference between them. **Passive filters** use only passive components (resistors, capacitors, and inductors) to create a frequency-dependent response, delivering an output which is at best equal in size to the input. **Active filters**, while they may also involve any of the passive components, normally use only resistors, capacitors, and one or more *active devices*, such as transistors or op amps. An active filter may provide the same frequency response as a passive filter but can also exhibit *gain*, provided by the active device, so the output may be *larger* than the input.

Figure 5.1 illustrates the basic "L," "T," and "pi" connection modes. The **"L" filter**, the simplest mode, contains one or more components in the *series* position and one or more in the *shunt* position, always located on the output side. The **"T" mode** consists of two series sections, one on each side of the shunt section. The **"pi" mode** consists of two shunt sections separated by a series section. In reviewing passive filters, the techniques discussed for identifying "L" filter types can also be applied to identification of the others.

Passive filters can be separated into two families by their component content. One family, **resonant** filters, includes band-pass and band-reject types. The second family, **nonresonant** filters, includes low-pass and high-pass

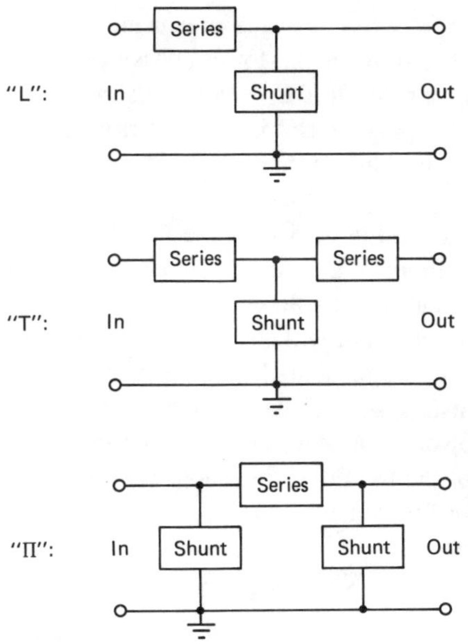

Figure 5.1

types. To identify an unknown filter, we first test to determine the family to which it belongs, then decide which of the two types of responses in its family the filter must provide.

To identify the family to which a filter belongs, look carefully at the circuit. All resonant filters must contain both a capacitor and an inductor, and both must be found in the same segment of the filter, together in either the series position or the shunt position. All resonant filters *must* meet these requirements. If the filter contains both inductance and capacitance, but they are in different segments of the filter, or if the filter contains only one reactive component, then the filter is nonresonant.

After determination of the family of a filter, an appropriate second test will determine its specific response type.

5.1 PASSIVE NONRESONANT FILTERS

Low-pass and high-pass filters must always contain at least one reactive component, since reactance varies with frequency, while resistance does not. If the filter contains only one reactive component, it is clearly nonresonant. If the filter contains both an inductor and a capacitor, but one is in series and the other in shunt, it is still a nonresonant filter.

A low-pass filter will block high frequencies at the output. A high-pass filter will block low frequencies but pass high frequencies at the output. The easiest way to tell if a nonresonant filter is a low-pass or high-pass is to consider its action for the lowest possible frequency, 0 Hz (DC). Consider how the network reacts to DC voltage at its input. If DC appears at the output, the filter is a low-pass. If DC is blocked from appearing at the output, the filter is a high-pass. For this purpose, remember that X_L is zero ohms (a short) for DC, while X_C is open, giving an infinite reactance value for DC.

Low-pass filters are shown in Figure 5.2. In the *RC* filter, DC at the input will charge the capacitor, developing a DC output. In the *RL* filter, the inductor is a short for DC, which places all input voltage across the resistor, and thus at the output. In the *LC* filter, the capacitor will charge through the inductor, again developing DC at the output. Although all are low-pass filters, since each circuit has its own reactance relationship, so each has its own formula for determination of the rolloff or half-power frequency, at which the output amplitude is .707 times the maximum. For the *RC* filter, this happens when $R = X_C$, which is then rearranged to solve for frequency:

$$R = \frac{1}{6.28 \cdot f \cdot C} \qquad (5.1)$$

$$f = \frac{1}{6.28 \cdot R \cdot C} \qquad (5.2)$$

EXAMPLE 5.1

In a low-pass *RC* filter, what is the characteristic rolloff or 3-dB frequency if:

(a) $R = 1\ \text{k}\Omega$ and $C = 1\ \mu\text{F}$?

$$f = \frac{1}{6.28 \cdot 1000 \cdot .000001} = 159 \text{ Hz}$$

(b) $R = 330\ \Omega$ and $C = .03\ \mu\text{F}$?

$$f = \frac{1}{6.28 \cdot 330 \cdot .00000003} = 16{,}076 \text{ Hz}$$

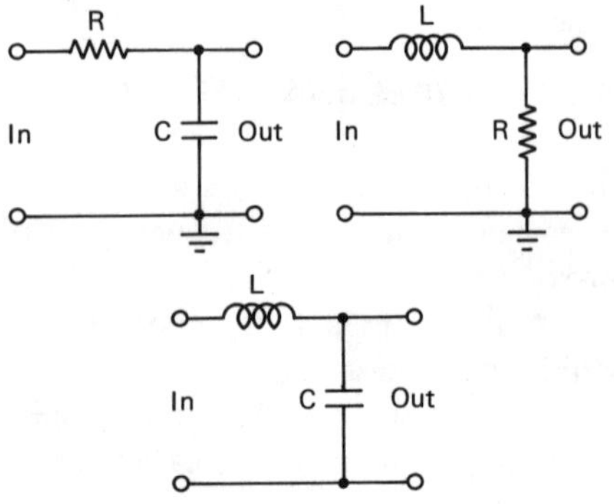

Figure 5.2

For the *RL* filter, the rolloff frequency happens when $R = X_L$:

$$R = 6.28 \cdot f \cdot L \tag{5.3}$$

$$f = \frac{R}{6.28 \cdot L} \tag{5.4}$$

EXAMPLE 5.2

In a low-pass *RL* filter, what is the characteristic rolloff or 3-dB frequency if:

(a) $R = 120\ \Omega$ and $L = 10$ mH?

$$f = \frac{120}{6.28 \cdot .01} = 1911 \text{ Hz}$$

(b) $R = 8.2\ \Omega$ and $L = 35$ mH?

$$f = \frac{8.2}{6.28 \cdot .035} = 37.3 \text{ Hz}$$

For the *LC* filter, the frequency occurs when $X_L = X_C$:

$$6.28 \cdot f \cdot L = \frac{1}{6.28 \cdot f \cdot C} \tag{5.5}$$

$$f = \frac{1}{6.28\sqrt{L \cdot C}} \tag{5.6}$$

EXAMPLE 5.3

In a low-pass *LC* filter, what is the characteristic rolloff or 3-dB frequency if:

(a) $L = 2$ mH and $C = 5\ \mu$F?

$$f = \frac{1}{6.28 \cdot \sqrt{.002 \cdot .000005}} = 1592 \text{ Hz}$$

(b) $L = 30$ mH and $C = 120\ \mu$F?

$$f = \frac{1}{6.28 \cdot \sqrt{.03 \cdot .00012}} = 83.9 \text{ Hz}$$

Notice that the formula for the *LC* filter is the same as the resonant-frequency formula, even though the filter is *not* actually resonant. In all formulas, the value of two times pi has been approximated as 6.28.

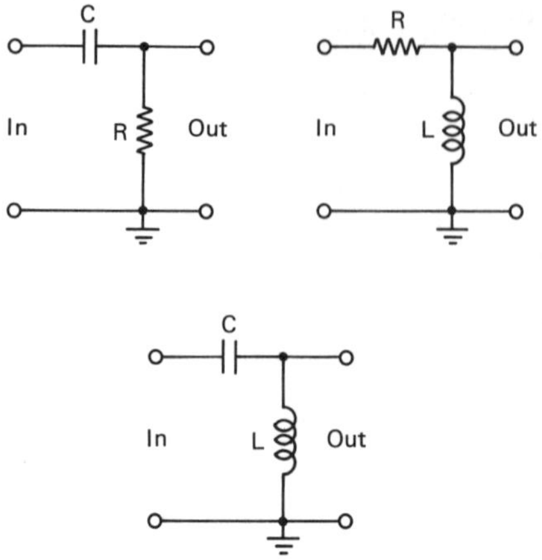

Figure 5.3

Passive high-pass filters are shown in Figure 5.3. Since the conditions for rolloff are identical to those for the low-pass filters, with resistance and reactance equal in the *RC* and *RL* filters, the rolloff frequency formulas for these filters are also equation (5.2) for the *RC* high-pass and equation (5.4) for the *RL* high-pass. In the *LC* high-pass, both reactances must be equal, which calls for equation (5.6). The only difference here is that in both the *RC* and *LC* filters, the presence of the capacitor in the series position will, once charged, block DC from the output. In the *RL* filter, the inductor in the shunt position will, with X_L being zero, short out any DC, so none appears at the output. Since DC is blocked from the output in these nonresonant filters, all are high-pass filters.

EXAMPLE 5.4

Find the characteristic rolloff or 3-dB frequency for each of these high-pass filters:

(a) $R = 15\ \text{k}\Omega$ and $C = .15\ \mu\text{F}$:

$$f = \frac{1}{6.28 \cdot 15{,}000 \cdot .00000015} = 70.7\ \text{Hz}$$

(b) $L = 5\ \text{mH}$ and $C = 4.7\ \mu\text{F}$:

$$f = \frac{1}{6.28 \cdot \sqrt{.005 \cdot .0000047}} = 1038\ \text{Hz}$$

(c) $R = 22$ ohms and $L = 125\ \text{mH}$:

$$f = \frac{22}{6.28 \cdot .125} = 28.0\ \text{Hz}$$

5.2 PASSIVE RESONANT FILTERS

The key to identification of a resonant filter, as previously specified, involves not only the presence of inductance and capacitance in the filter, but also their occurrence together. If they are not together in the same section, the filter is not resonant. Once you have identified a filter as resonant, the next procedure is to think of X_{total} for the combination of reactances present: if in series, X_{total} becomes zero ohms, or a short, for the resonant frequency; if in parallel, X_{total} becomes infinite, or an open, at resonance. Determine how the filter will react to the resonant frequency at its input: if it is blocked from reaching the output, the filter is a band-reject; if it reaches the output easily, while other frequencies experience attenuation, the filter is a band-pass.

All resonant filters use the resonant-frequency formula, equation (5.6), to determine the center frequency of the pass or reject band. Resistance, when present in the filter, determines the width of the pass or reject band by determining quality factor, bandwidth, and half-power frequencies for the filter. The resistance plays *no* part, however, in setting the center frequency, which is determined *solely* by the inductance and capacitance values.

Band-pass filters are shown in Figure 5.4. The top filter, with a series combination (X_{total} goes to 0) in the series position, will easily pass the resonant frequency to the output, but will develop a net reactance at other frequencies, so that the output will be reduced. In the lower filter, a parallel combination (X_{total} goes to infinity) appears in the shunt position. Since impedance is maximum at resonance for a parallel circuit, this too will easily pass the resonant frequency to the output, while other frequencies provide reduced amplitudes.

Band-reject filters are shown in Figure 5.5. The top filter shows a parallel combination (X_{total} goes to infinity) in the series position, which, with maximum impedance at the resonant frequency, blocks it from passing through. In the

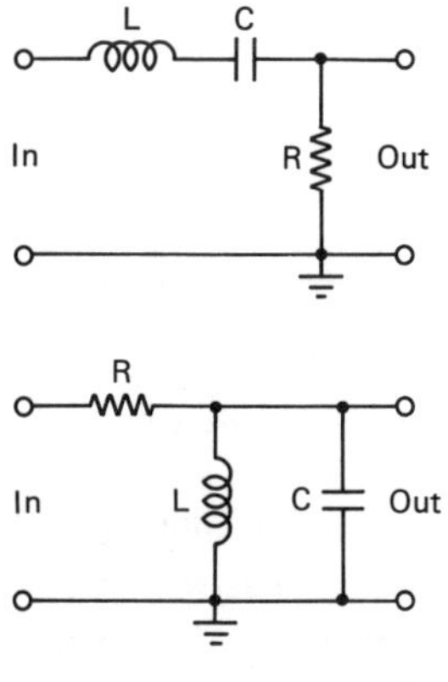

Figure 5.4

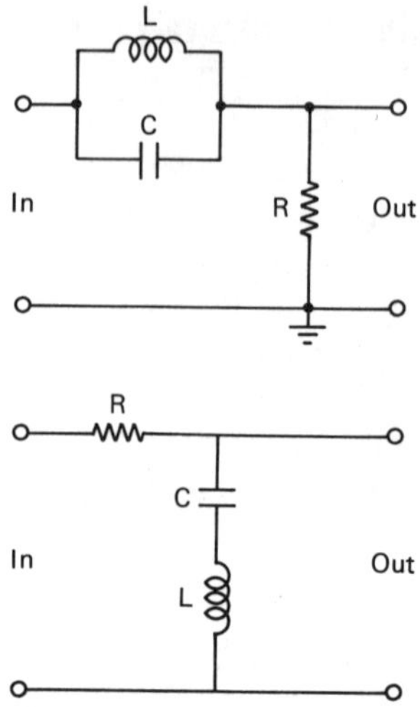

Figure 5.5

lower filter, a series combination (X_{total} goes to 0) appears in the shunt position, effectively shorting the resonant frequency to ground.

EXAMPLE 5.5

Find each center frequency for:

(a) Band-pass filter with $C = .05\ \mu F$, $L = 300\ \mu H$, and $R = 12\ k\Omega$:

$$f = \frac{1}{6.28\sqrt{.0003 \cdot .00000005}} = 41{,}094 \text{ Hz}$$

(b) Band-reject filter with $L = 4$ mH, R = 820 Ω and $C = 15\ \mu F$:

$$f = \frac{1}{6.28\sqrt{.004 \cdot .000015}} = 650 \text{ Hz}$$

(c) Band-pass filter with $R = 120\ k\Omega$, $C = .002\ \mu F$, and $L = 50\ \mu H$:

$$f = \frac{1}{6.28\sqrt{.00005 \cdot .000000002}} = 503{,}292 \text{ Hz}$$

5.3 ACTIVE LOW-PASS AND HIGH-PASS FILTERS

As we have already seen, a passive low-pass or high-pass filter can be constructed with only two components. An active filter will include a transistor or an op amp, so it will be more expensive to build, and it will also be physically somewhat larger than the passive filter. Despite these facts, active filters are

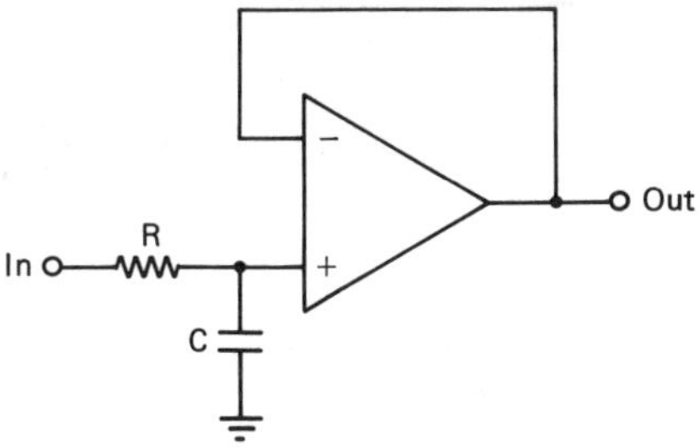

Figure 5.6

used in many applications where passive filters were used previously. The main reason is that the active filter can provide the same frequency response as the passive filter but in addition can provide gain throughout the band of frequencies that it passes, a feature that passive filters cannot duplicate. In the selection of feedback component values, the frequency-response and gain characteristics of an active filter can be tailored to satisfy virtually any requirement. In addition, some active filter circuits provide a steeper rolloff rate for frequencies which are outside the passband than can be achieved with any passive filter.

The simplest variety of active low-pass filter is shown in Figure 5.6. The output is directly connected to the inverting input, so the op amp is set for unity gain. If a larger gain is desired, the short from output to inverting input can be replaced with a gain-determining voltage divider. In actuality, a simple passive filter circuit appears on the input, so the op amp simply amplifies the signals that actually reach its true input. The rolloff frequency for this filter can be found using equation (5.2).

A more complex active low-pass filter is shown in Figure 5.7. This circuit will provide a rolloff rate which is twice as steep as the filter in Figure 5.6, so high frequencies above the passband are attenuated more thoroughly. All four passive components are involved with the rolloff frequency of this circuit:

$$f = \frac{1}{6.28\sqrt{R_1 \cdot R_2 \cdot C_1 \cdot C_2}} \tag{5.7}$$

EXAMPLE 5.6

In Figure 5.7, if R_1 is 10 kΩ, R_2 is 15 kΩ, C_1 is .05 μF, and C_2 is .1 μF, what is the rolloff frequency?

$$f = \frac{1}{6.28\sqrt{10{,}000 \cdot 15{,}000 \cdot .00000005 \cdot .0000001}} = 184 \text{ Hz}$$

For the filter in Figure 5.7, if R_1 and R_2 are the same resistance value, and C_1 and C_2 are the same capacitance value, the frequency equation simplifies to the form of equation (5.2).

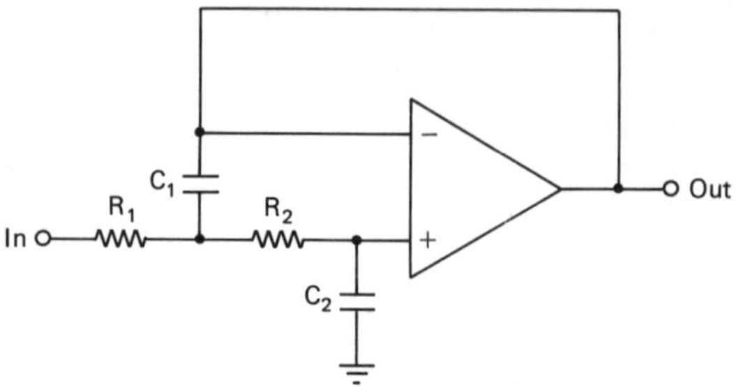

Figure 5.7

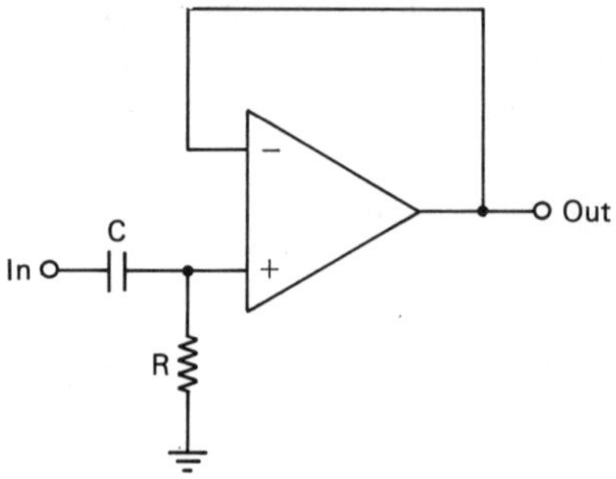

Figure 5.8

The simplest active high-pass filter circuit is shown in Figure 5.8. This op amp, like the one in Figure 5.6, is set for unity gain, but the short from output to inverting input can be replaced with a gain-determining voltage divider if a gain of more than 1 is required. The rolloff frequency for this circuit can also be determined by using equation (5.2). Like the simple low-pass filter in Figure 5.6, the rolloff occurs in the input circuit, so this op amp, too, simply amplifies the signals that arrive at its true input.

A more complex active high-pass filter appears in Figure 5.9. This circuit will provide a rolloff rate which is twice as steep as the simple circuit in Figure 5.8, so this filter will reject low frequencies more thoroughly. To find the rolloff frequency for this filter circuit, equation (5.7), seen with the low-pass circuit, applies for this high-pass format as well. As with the low-pass, this formula

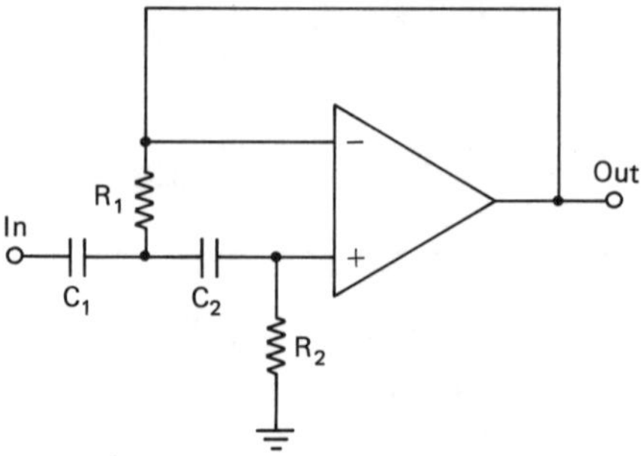

Figure 5.9

must be used when either R_1 and R_2, or C_1 and C_2 are different. However, if $R_1 = R_2$ and $C_1 = C_2$, then equation (5.2) can be used instead.

EXAMPLE 5.7

In Figure 5.9, if R_1 and R_2 are 2 kΩ, and C_1 and C_2 are .1 μF, what is the rolloff or 3-dB frequency?

$$f = \frac{1}{6.28\sqrt{2000 \cdot 2000 \cdot .0000001 \cdot .0000001}}$$

$$= \frac{1}{6.28 \cdot 2000 \cdot .0000001} = 796 \text{ Hz}$$

5.4 ACTIVE BAND-PASS AND BAND-REJECT FILTERS

Passive filters belonging to the resonant filter family contain a combination of inductance and capacitance, with the two components found together in the same section of the filter to provide resonant action. Active filters of this family, however, often contain only resistors and capacitors, arranged in such a way that two separate but interactive feedback paths within the circuit reinforce each other, but only within a very limited frequency range for band-pass filters, while both feedback paths will oppose each other within a limited frequency range for band-reject filters. Active filters containing a resonant combination, while offering the tunability and bandwidth control associated with an LC circuit, have been almost completely superseded by active *RC*-based resonant filters, which avoid the imperfections of inductors completely by not using them.

Figure 5.10 shows an active band-pass filter which uses a parallel tank circuit in the feedback path to determine its frequency sensitivity. Since a parallel tank circuit presents its maximum impedance level at the resonant frequency, this frequency will be amplified to a greater degree than any other frequency,

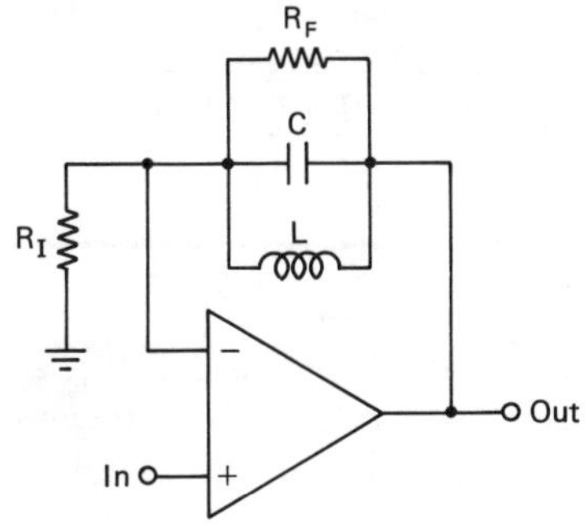

Figure 5.10

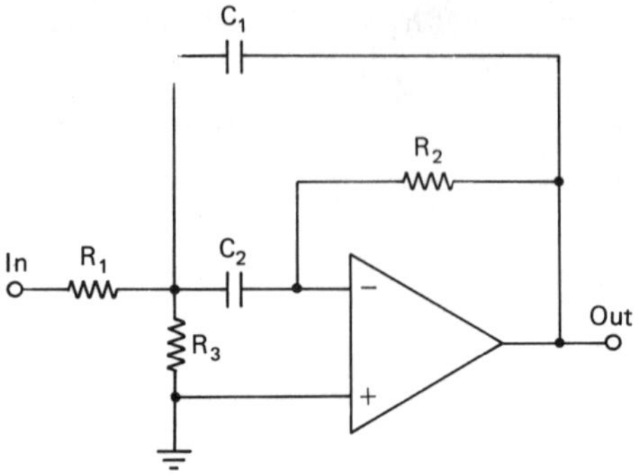

Figure 5.11

for which the impedance in the feedback path will be smaller, making the gain smaller as well. Resonant active filters generally place the resonant circuit in the feedback path, as shown in Figure 5.10. The use of a series combination in the feedback loop would present its minimum impedance level at the resonant frequency, resulting in the gain becoming very small at or near the resonant frequency, so that the filter circuit provides a band-reject response.

An *RC*-based active band-pass filter is shown in Figure 5.11. Note that two feedback loops are present: one gives a low-pass response, the other a high-pass response. Both feedback paths are superimposed to provide a band-pass response. The center frequency of this filter's pass band can be determined by the following formula:

$$f = \frac{1}{6.28\sqrt{\dfrac{R_1 \cdot R_2 \cdot R_3 \cdot C_1 \cdot C_2}{R_1 + R_3}}} \tag{5.8}$$

EXAMPLE 5.8

In Figure 5.11, if C_1 and C_2 are both .01 μF, and all three resistors are 10 kΩ, what is the center frequency?

$$f = \frac{1}{6.28\sqrt{\dfrac{10\ \text{k}\Omega \cdot 10\ \text{k}\Omega \cdot 10\ \text{k}\Omega \cdot .01\ \mu\text{F} \cdot .01\ \mu\text{F}}{10\ \text{k}\Omega + 10\ \text{k}\Omega}}} = 2252\ \text{Hz}$$

An *RC*-based active band-reject filter is shown in Figure 5.12. Note that this circuit, quite similar to the circuit in Figure 5.11, also provides two feedback loops: one low-pass, and the other high-pass. In this circuit, both feedback

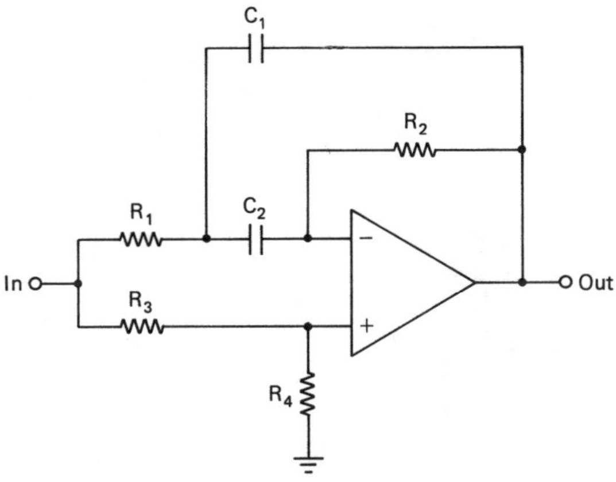

Figure 5.12

paths combine to provide a band-reject response. The formula for determination of the center frequency of the reject band is equation (5.7).

EXAMPLE 5.9

In Figure 5.12, if C_1 and C_2 are .02 and .03 μF, and all resistors are 22 kΩ, what is the center frequency?

$$f = \frac{1}{6.28\sqrt{22\text{ k}\Omega \cdot 22\text{ k}\Omega \cdot .02\ \mu\text{F}\ .03\ \mu\text{F}}} = 295\text{ Hz}$$

Another interesting way of achieving an active band-pass response is to cascade a low-pass with a high-pass filter, as shown in Figure 5.13. In most cases, the high-pass filter has a rolloff frequency somewhat lower than the low-pass cutoff frequency, resulting in a band of frequencies between the two which both stages will pass fairly well. Through the use of various separations between the two frequencies, as well as different rolloff rates, band-pass type active filters can be constructed to duplicate the frequency responses of many resonant filters. In addition, response patterns which are not available from resonant filters can be produced by cascaded low-pass and high-pass combinations.

Active band-reject responses can also be achieved by combining low-pass and high-pass filters so that their outputs are added, as shown in Figure 5.14. With this arrangement, the low-pass rolloff frequency is below the high-pass rolloff frequency. With both outputs added, frequencies below the low rolloff will be supplied by the low-pass filter, and frequencies above the high rolloff will be supplied by the high-pass filter. Only frequencies between the two rolloff frequencies, since neither filter will pass them, are rejected by this band-reject circuit.

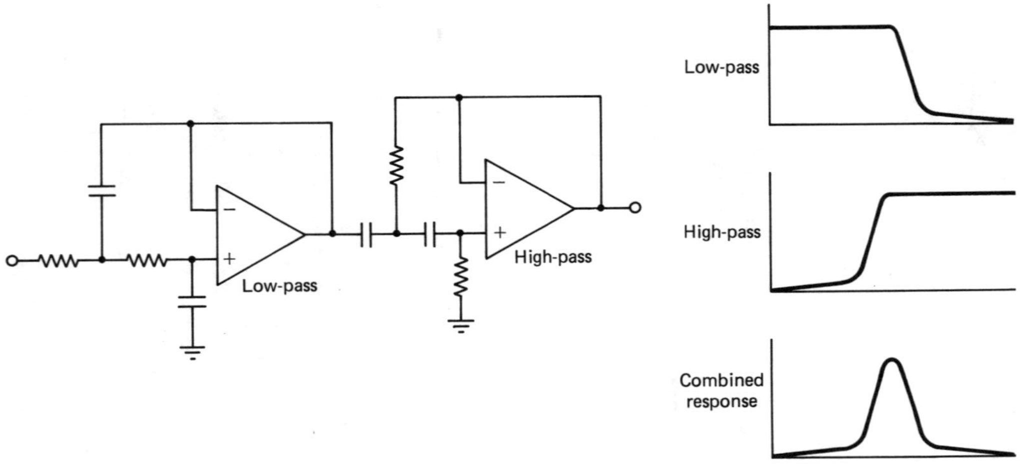

Figure 5.13

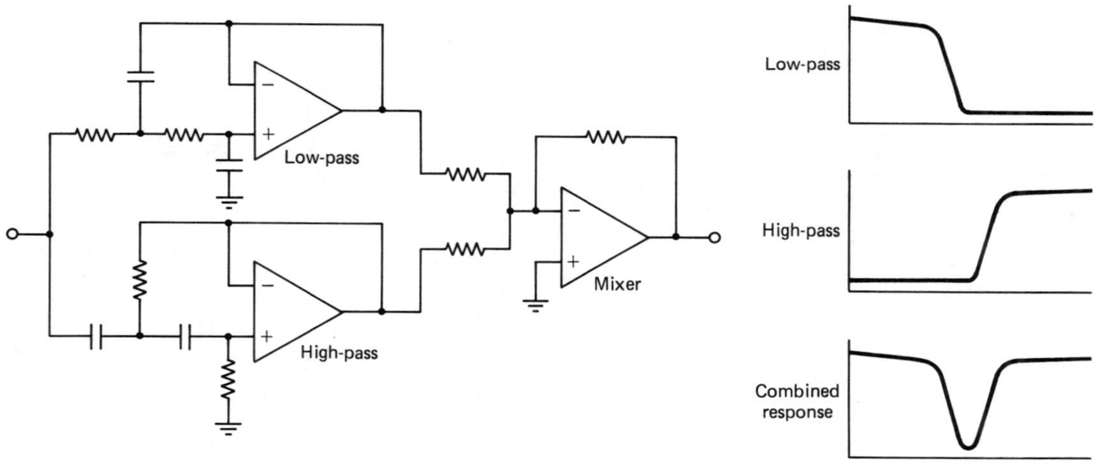
Low-pass
High-pass
Mixer
Low-pass
High-pass
Combined
response

Figure 5.14

5.5 TROUBLESHOOTING ACTIVE FILTER CIRCUITS

When a problem arises in an active filter circuit, one of three things will occur: no output is seen, the output is present but distorted, or output amplitudes indicate incorrect frequency response. If no output appears, first check DC supply voltages and input and output terminal voltages on the op amp. If proper supply voltage is present and inputs and output are at ground with dual supplies, or half the supply value for a single supply, three possibilities remain. If the circuit has a series capacitor in the input area, it could be defective. A defective solder connection could be at fault. The op amp, of course, could also be defective.

When the output shows distortion, DC supplies and input and output bias voltages should be checked. An offset problem may have developed. Further possibilities include a capacitor with excessive leakage, a problem with one DC supply, or a defective op amp.

When the frequency response has changed, the op amp is probably not to blame. Potential causes include a defective capacitor, a defective resistor, or a bad solder connection. The most likely cause of a frequency-response change is a defective capacitor.

PROBLEMS

5.1. Refer to Figure 5.15 to identify each filter circuit (as low-pass, high-pass, band-pass, or band-reject) and determine each filter's rolloff or center frequency.

	Filter Type	3-dB or Center Frequency		Filter Type	3-dB or Center Frequency
(a)	______	__________	**(g)**	______	__________
(b)	______	__________	**(h)**	______	__________
(c)	______	__________	**(i)**	______	__________
(d)	______	__________	**(j)**	______	__________
(e)	______	__________	**(k)**	______	__________
(f)	______	__________	**(l)**	______	__________

5.2. Refer to Figure 5.16 to identify each active filter as band-pass or band-reject, and determine each filter's center frequency.

	Filter Type	3-dB or Center Frequency		Filter Type	3-dB or Center Frequency
(a)	______	__________	**(c)**	______	__________
(b)	______	__________	**(d)**	______	__________

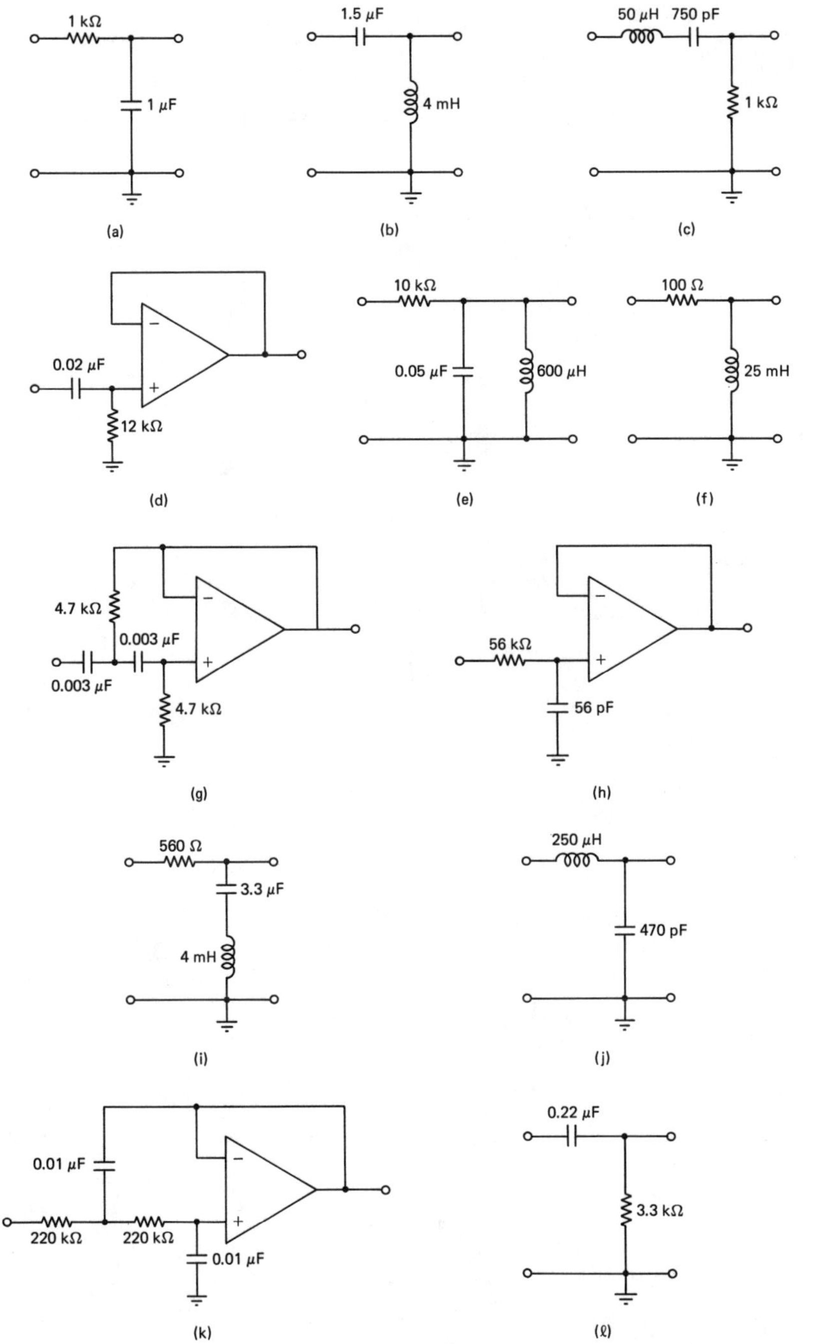
1 kΩ
1 μF
(a)
1.5 μF
4 mH
(b)
50 μH 750 pF
1 kΩ
(c)
0.02 μF
12 kΩ
(d)
10 kΩ
0.05 μF
600 μH
(e)
100 Ω
25 mH
(f)
4.7 kΩ
0.003 μF
0.003 μF
4.7 kΩ
(g)
56 kΩ
56 pF
(h)
560 Ω
3.3 μF
4 mH
(i)
250 μH
470 pF
(j)
0.01 μF
220 kΩ
220 kΩ
0.01 μF
(k)
0.22 μF
3.3 kΩ
(ℓ)

Figure 5.15

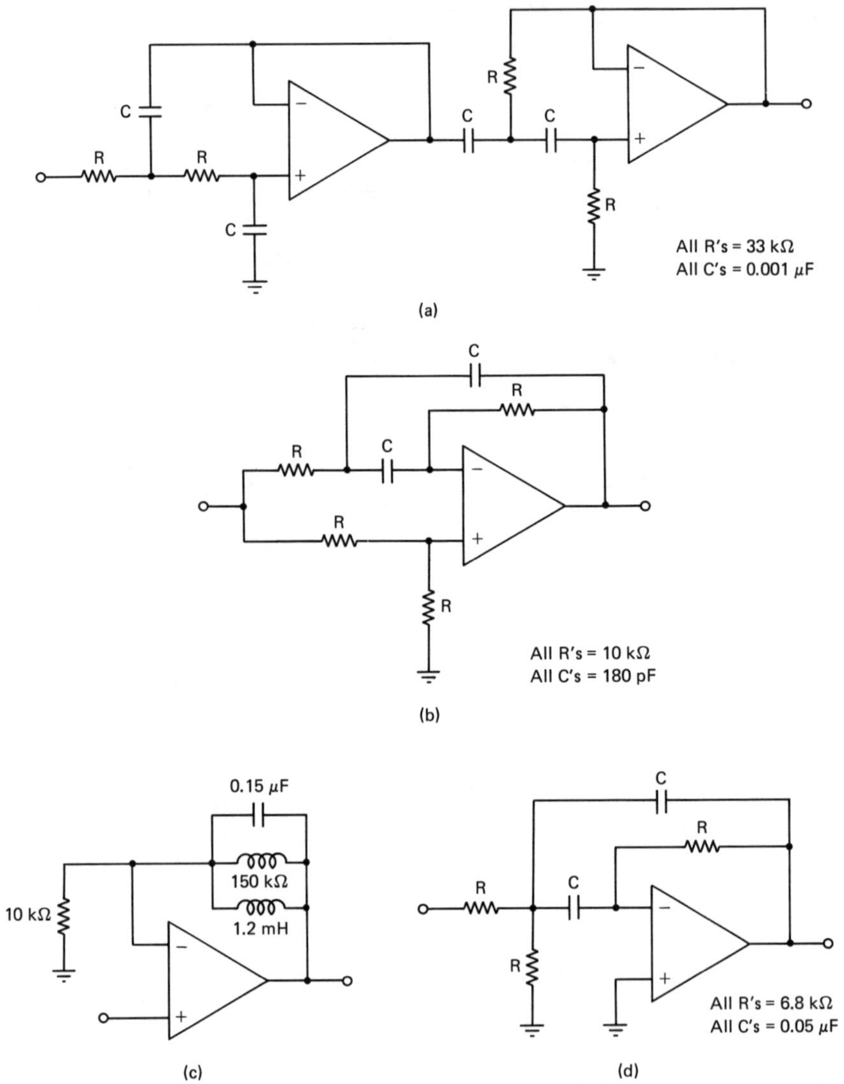
C
R
R
C
R
C
C
R
All R's = 33 kΩ
All C's = 0.001 μF
(a)
C
R
R
C
R
R
All R's = 10 kΩ
All C's = 180 pF
(b)
0.15 μF
150 kΩ
1.2 mH
10 kΩ
(c)
C
R
R
C
R
All R's = 6.8 kΩ
All C's = 0.05 μF
(d)

Figure 5.16

6 Sinusoidal Oscillator Circuits

An oscillator is a circuit which receives only one input: its DC supply voltage. Through the application of positive feedback which is also frequency-dependent, the oscillator circuit will provide an output signal of predictable amplitude and frequency. Unless they are damaged or incorrectly constructed, all of the oscillator circuits that we shall examine in this chapter should provide sinusoidal outputs. In some cases, the sine wave is generated by a resonant tank circuit; in others, it arises from *RC* networks. When an *LC* circuit generates the sine wave, its frequency of oscillation can be calculated using the standard resonant-frequency formula:

$$f = \frac{1}{6.28\sqrt{L \cdot C}} \tag{6.1}$$

For all oscillators based on *LC* resonance, this is the appropriate formula to be used. For nonresonant oscillators, each applicable oscillation-frequency formula will appear within its appropriate text section.

6.1 THE WEIN-BRIDGE OSCILLATOR

The **Wein-bridge oscillator** is an *RC*-based circuit, containing a special configuration of two resistors and two capacitors called a **lead-lag network.** As shown in Figure 6.1, a series *RC* combination is connected between the output and the noninverting input, while a parallel *RC* combination is connected between the noninverting input and ground. The series capacitor will, with its large reactance, tend to block low frequencies. The parallel capacitor will, with its small reactance, act to "short out" high frequencies. Only a small range of frequencies will be effectively passed by both networks, and the frequency which passes best of all, providing the maximum feedback amplitude, will be the frequency at which the circuit will oscillate.

A variation of the usual voltage divider is found between the output and the inverting input, setting the gain of the op amp. The Wein-bridge oscillator circuit requires a higher-than-normal gain when it is initially energized to begin oscillation. The initially tiny AC variation will grow in amplitude due to the high initial gain, until it reaches a sufficient amplitude and triggers a decrease of the gain to the normal operating level. Once this occurs, the oscillation amplitude stops increasing, stabilizing at its normal operating amplitude.

One of two unusual features is normally seen in the resistive voltage divider to provide this gain control. One way to accomplish this involves the use of a small light bulb in the position of the lower feedback resistor (R_i). When it is cold, a light bulb offers a low resistance, for a very large initial gain figure. As oscillation begins, the light bulb heats up, causing its resistance to increase, thus dropping the gain to a normal level. The second method involves the use of two series resistors for R_f, with two series-connected opposing zener diodes placed in parallel with one of the resistors. When the circuit is energized, the output size will build until it reaches sufficient size to "turn on" the diodes. The diodes then serve to short out the resistor with which they are in parallel, thus decreasing the gain to a normal level determined by the second feedback resistor.

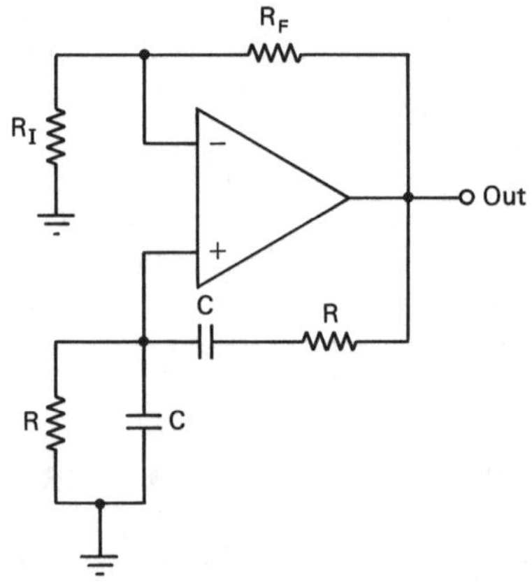

Figure 6.1

The Wein-bridge oscillator generally uses an op amp as its active device, and its frequency can be made variable by using a pair of variable capacitors in the lead-lag network, ganged so that they vary together. The frequency of oscillation for this circuit, with identical resistors and identical capacitors in the lead-lag network, is:

$$f = \frac{1}{6.28 \cdot R \cdot C} \tag{6.2}$$

EXAMPLE 6.1

If the oscillator in Figure 6.1 has .33 μF for both C's and 2.2 kΩ for both R's, what is the frequency of oscillation?

$$f = \frac{1}{6.28 \cdot 2200 \cdot .00000033} = 219 \text{ Hz}$$

6.2 THE PHASE-SHIFT OSCILLATOR

The **phase-shift oscillator,** shown in Figure 6.2 in both op amp and transistor versions, also uses RC networks to set the oscillation frequency. With a transistor, the common-emitter format is used, owing to its 180-degree phase shift between output and input. With an op amp, the network is connected between the output and the inverting input, providing the same 180-degree phase shift.

Three series RC high-pass networks comprise the entire feedback path. Since equal capacitor values are almost invariably used, each RC network must provide a 60-degree phase shift, for a total shift of 180 degrees through all three networks. The phase-shift network causes a significant attenuation of the signal between the output and the feedback point, so a rather large gain factor (at least 29) must be present in the amplifying device to compensate for this loss. Assuming all three capacitors are the same size, and their associated resistors are all the same size, the formula for finding this circuit's oscillation frequency takes a rather unique form owing to its unique 60-degree shift in each RC network:

$$f = \frac{1}{6.28\sqrt{6} \cdot R \cdot C} = \frac{1}{15.39 \cdot R \cdot C} \tag{6.3}$$

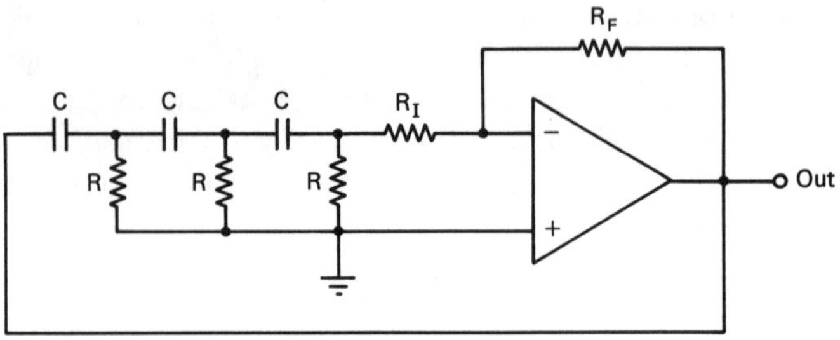

(a)

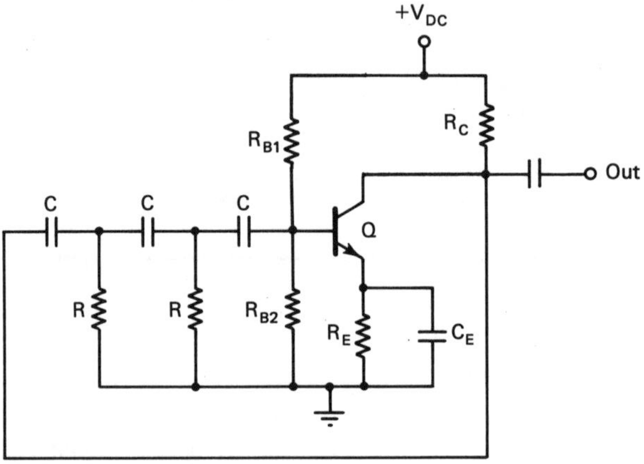

(b)

Figure 6.2

EXAMPLE 6.2

If the circuit in Figure 6.2(a) uses three 750-pF capacitors and three 82-kΩ resistors, what is the frequency of oscillation?

$$f = \frac{1}{15.39 \cdot 82{,}000 \cdot .00000000075} = 1057 \text{ Hz}$$

The phase-shift oscillator, like the Wein-bridge oscillator, can also be constructed to permit adjustability of its oscillation frequency. With three capacitors in the feedback network, the adjustable phase-shift oscillator will require a ganged set of three capacitors, which are varied together when the frequency is adjusted.

6.3 THE HARTLEY OSCILLATOR

The **Hartley oscillator** circuit is based on resonance in an *LC* tank for frequency determination. The inductor in the tank is actually composed of either two separate series inductors, or a single coil with a tap, with the shared connection or tap connected to ground in the op amp circuit, and to the DC supply in the transistor circuit. By holding this shared point at a constant voltage level, a variation in voltage at one end of the tank circuit will cause an opposite type of variation, 180 degrees out of phase, at the other end. The 180-degree phase shift through the tank circuit makes this format suitable for both the inverting op amp and the common-emitter transistor amplifier. Figure 6.3 illustrates both the op amp and transistor versions of the Hartley oscillator. Note that the transistor circuit contains a coupling capacitor in series between the base and the tank circuit. With the transistor, unlike the op amp, bias conditions require DC isolation for the base terminal, easily achieved with a coupling capacitor.

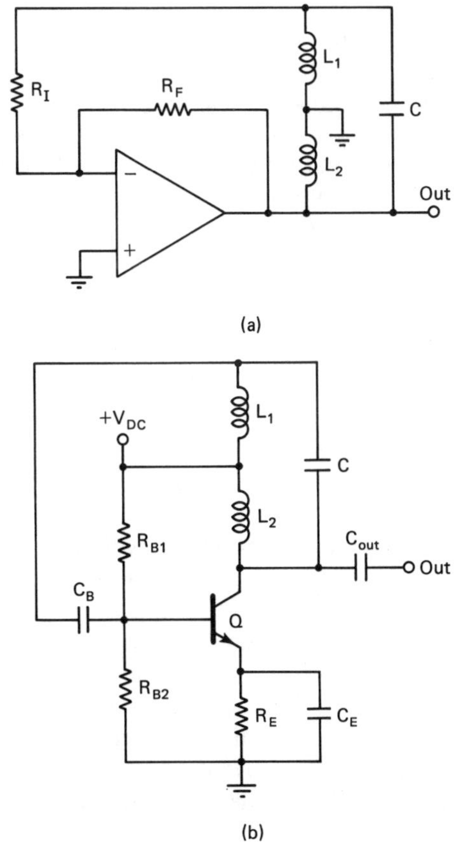

Figure 6.3

To determine the resonant frequency of the tank, both series inductance values must be added. The frequency of resonance is then determined by using the total inductance and capacitance values in equation (6.1), the standard resonant-frequency formula.

EXAMPLE 6.3

If the circuit in Figure 6.3(a) has 2.2 μF for C, 20 μH for L_1, and 120 μH for L_2, what is the oscillation frequency?

$$L_{\text{total}} = L_1 + L_2 = 140\ \mu\text{H}$$

$$f = \frac{1}{6.28\sqrt{.00014 \cdot .0000022}} = 9073\ \text{Hz}$$

Note that in the transistor circuit, the frequency is determined solely by the parallel tank circuit and is not affected by the value of the coupling capacitor, which simply couples the AC signal back to the base without disturbing DC bias conditions.

The Hartley oscillator can be made to provide a variable frequency of oscillation by using a variable capacitor in the tank circuit to alter its resonant frequency. A large range of frequencies can be achieved by using a rotary switch connected to various values of capacitance.

6.4 THE COLPITTS OSCILLATOR

The **Colpitts oscillator,** like the Hartley, is also based on LC resonance. In the Colpitts, the tank circuit involves the presence of two series capacitors, rather than two series inductors. The end result, with the capacitors' shared node connected to the DC supply or ground, is a 180-degree phase shift between the two ends of the tank, making this circuit suitable for use with either the inverting op amp or the common-emitter transistor circuit. Figure 6.4 illustrates both the op amp and transistor versions.

Figure 6.5 shows a different Colpitts transistor oscillator which uses a common-base format! In this version, the feedback is provided in phase with the output waveform. This is the result of connecting one end of the tank to the supply voltage, rather than the shared terminal, so the other two tank connections will be in phase with each other. With its ability to provide voltage gain from an in-phase relationship, the common-base circuit is suitable for this application. Note that with the transistor, the coupling capacitor, necessary in the

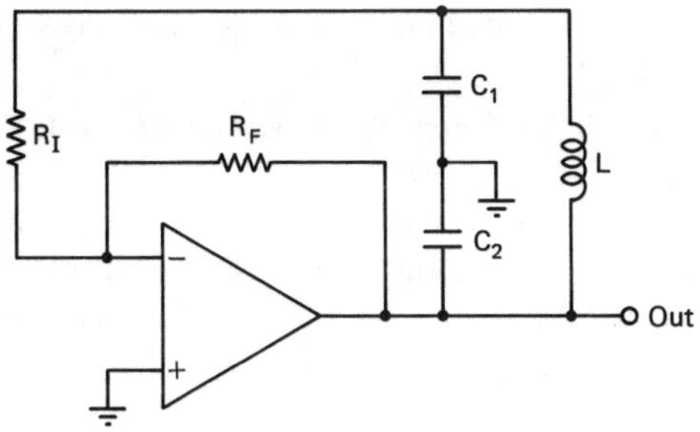

(a)

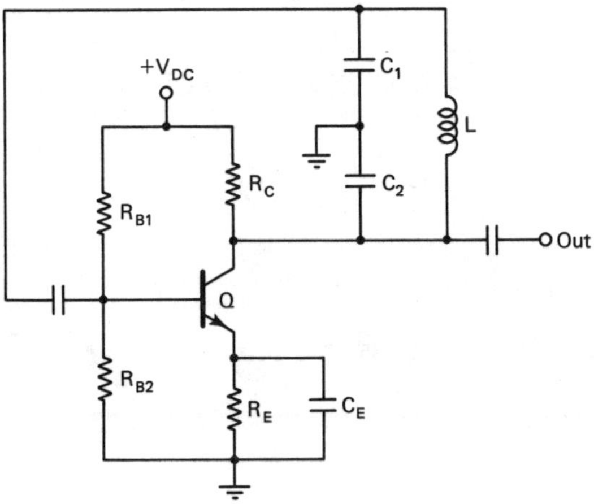

(b)

Figure 6.4

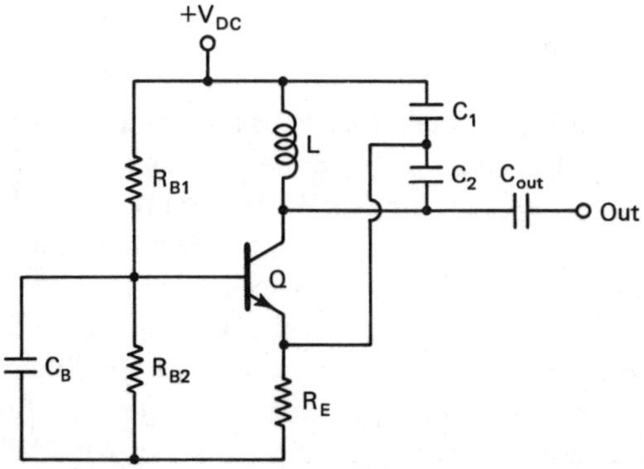

Figure 6.5

Hartley, is not required in the Colpitts, since the two tank capacitors can block DC levels while oscillating.

To find the resonant frequency of any Colpitts format, the two capacitors must first be combined. Since they are in series, product over sum must be used, resulting in an equivalent capacitance value which will be smaller than either series capacitance value. Once the single equivalent capacitance has been found, it and the parallel inductance value determine the resonant frequency through equation (6.1), the standard resonant-frequency formula.

EXAMPLE 6.4

If the circuit in Figure 6.4 has 60 pF for C_1, 300 pF for C_2, and 120 μH for L, what is the resonant frequency?

$$C_{\text{total}} = \frac{C_1 \cdot C_2}{C_1 + C_2} = 50 \text{ pF}$$

$$f = \frac{1}{6.28\sqrt{.00012 \cdot .00000000005}}$$

$$= 2{,}055{,}724 \text{ Hz}$$

The Colpitts oscillator can be made to provide an adjustable frequency of oscillation by using a variable inductance in the construction of the tank circuit.

6.5 THE CLAPP OSCILLATOR

Like the Colpitts, the **Clapp oscillator** also uses two series capacitors, grounded at the shared terminal in order to force the two ends of the tank circuit to be 180 degrees out of phase. The primary difference between the two formats is the presence of one additional capacitor in the Clapp's tank circuit. As shown in Figure 6.6, this extra capacitance is placed in series with the inductor. The two series capacitors on the other side of the tank circuit maintain proper phase and amplitude relationships, just as they do in the Colpitts oscillator, but are no longer directly involved in determining the frequency of oscillation in this circuit, which is determined by the capacitor in series with the inductor through equation (6.1), the standard resonance formula.

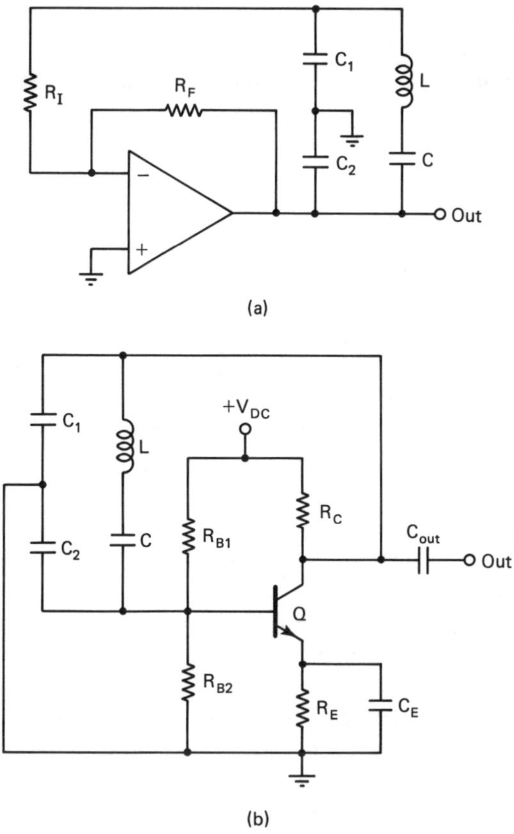

Figure 6.6

EXAMPLE 6.5

If the circuit in Figure 6.6(a) contains a 3-mH inductor, with a .15-μF capacitor in series, what is the frequency of oscillation?

$$f = \frac{1}{6.28\sqrt{.003 \cdot .00000015}} = 7506 \text{ Hz}$$

Since only one inductor and capacitor are directly involved in the resonant tank, the Clapp oscillator offers the best stability of oscillation frequency available among *LC*-based oscillators.

6.6 CRYSTAL OSCILLATORS

RC-based oscillators do not provide a completely reliable oscillation frequency. A change in temperature can cause variations in both resistor and capacitor values, which will lead to a change in the oscillation frequency. *LC*-based oscillators offer somewhat better frequency stability, but they, too, can be affected by temperature changes, as well as physical shock or vibration, movement of metal near the tank circuit, or a change in the *DC* supply voltage, again leading to changes in the oscillation frequency.

For maximum stability and predictability of the frequency of oscillation, an oscillator containing a **crystal** is the best choice. A quartz crystal is a **piezoelectric** material, which means it can transfer energy in either direction between vibrational and electrical forms. Thus, when it vibrates, a voltage is produced. And, when an AC voltage is applied to the crystal, it can begin to vibrate. *The frequency of vibration is determined solely by the physical dimensions of the crystal itself.* A crystal is usually enclosed in a conductive shell, which not only shields the crystal from external interference, but also provides a location for marking the crystal's resonant frequency, which may be specified to an accuracy of six digits or more.

Through its piezoelectric capabilities, the crystal is capable of replacing an entire *LC* tank circuit, setting the oscillator frequency with *its own internal frequency* of vibration. In this series mode, the reactance of the crystal will be smallest at its resonant frequency. When used in the feedback path, the crystal will therefore provide maximum feedback at its own frequency of resonance. Figure 6.7 shows both transistor and op amp examples of series-resonant crystal oscillators.

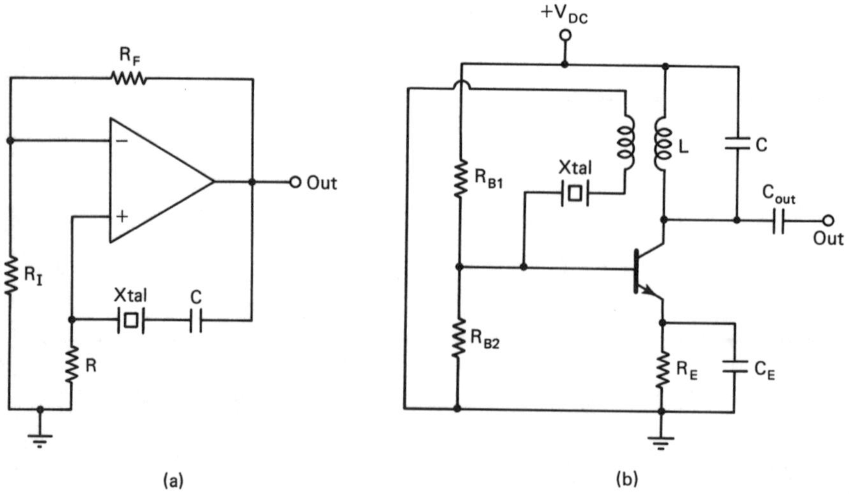

Figure 6.7

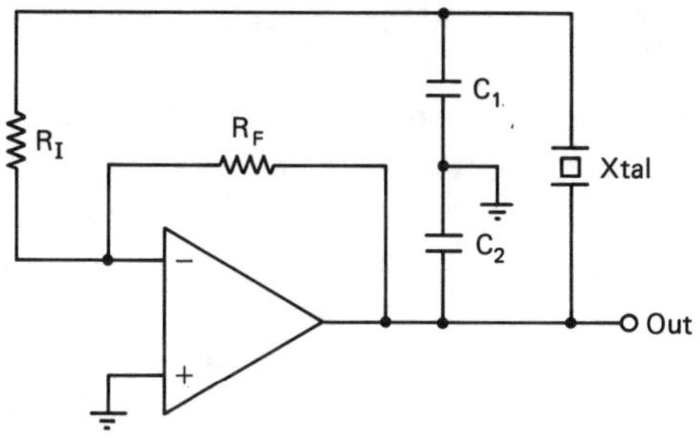

(a)

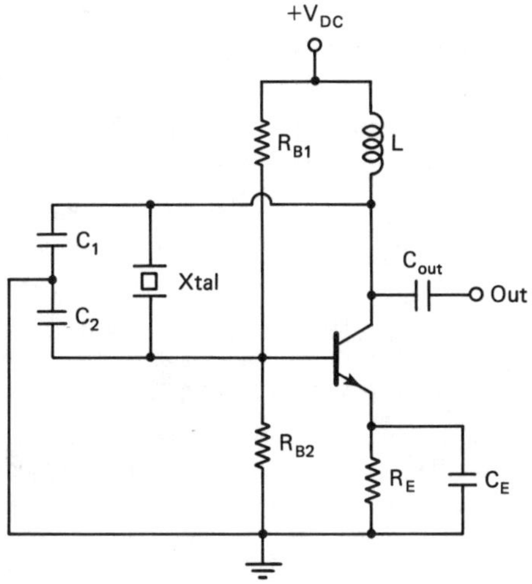

(b)

Figure 6.8

When it is placed in parallel with a capacitor, the crystal will exhibit parallel properties, with its impedance rising to maximum at the resonant frequency. Figure 6.8 illustrates parallel crystal resonance in both transistor and op amp examples, where the parallel crystal configuration is used to develop maximum voltage amplitude at the frequency of resonance.

Crystals do have certain limitations. One problem is that a crystal's resonant frequency will change with temperature, although far less change occurs than with *RC* or *LC* circuits. When precise control of the oscillation frequency is necessary, the crystal, and sometimes the entire oscillator circuit, will be placed in an "oven" in which the temperature will be closely controlled. By keeping its temperature as steady as possible, the crystal's frequency of oscillation will also be as steady as possible.

6.7 TROUBLESHOOTING OSCILLATORS

If no output appears, check for proper DC supply voltages. Verify that they actually reach the integrated circuit. If supplies are working, check the input and output terminals for DC levels: they should be at ground with dual +/− supplies, or at approximately half the supply voltage with a single supply. Readjust offset if necessary. If all DC levels appear normal, carefully inspect the auxiliary components around the op amp for defects. Finally, the op amp itself may be defective.

If a distorted output signal is present, check DC supplies and op amp terminal voltages as usual. If one supply is absent or low, distortion is likely to occur. Readjust offset if necessary. Next, check waveforms in the feedback circuitry and at the input terminal, in case a component or connection is defective. If all else looks right, the op amp itself is probably defective.

If the output provides a different frequency than it should, the op amp is probably not at fault. In this situation, carefully check all reactive components (capacitors and inductors) for noticeable exterior damage, and check all solder connections. A capacitor with an unusually low leakage resistance can easily cause a change in frequency.

PROBLEMS

6.1. In any functional oscillator circuit, what two special properties, responsible for oscillator operation, are always required in the feedback circuit?

____________________ and ____________________

6.2. The Hartley, Colpitts, and Clapp oscillators, being resonant in nature, all use the same frequency formula. What other special feature is seen in the tank circuits in all three of these formats?

____

6.3. What types of events can cause an oscillator circuit to change its frequency of operation, either temporarily or permanently? List as many causes as you can:

____

____

____

____

____

____

____

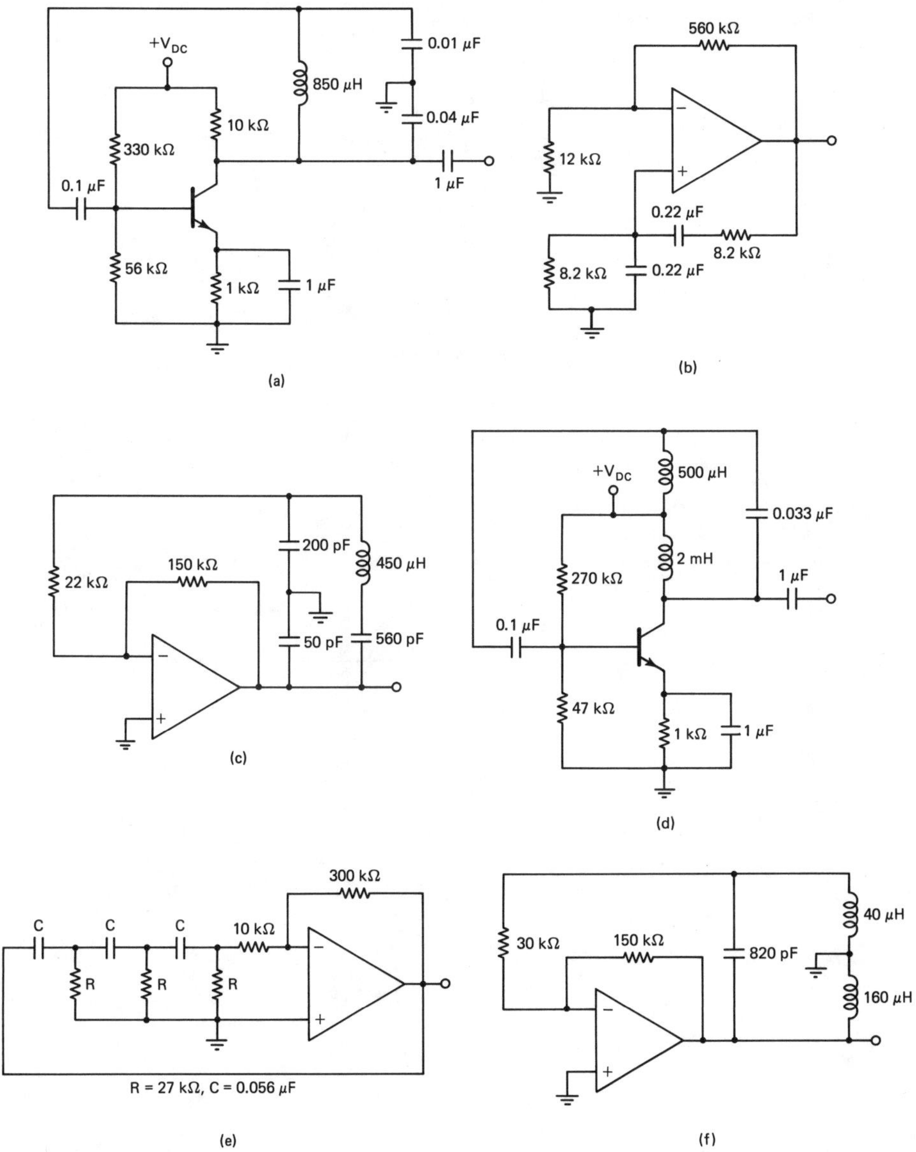
+V_DC
850 μH
0.01 μF
0.04 μF
10 kΩ
330 kΩ
0.1 μF
1 μF
56 kΩ
1 kΩ
1 μF
(a)
560 kΩ
12 kΩ
0.22 μF
8.2 kΩ
8.2 kΩ
0.22 μF
(b)
22 kΩ
150 kΩ
200 pF
450 μH
50 pF
560 pF
(c)
+V_DC
500 μH
0.033 μF
2 mH
270 kΩ
1 μF
0.1 μF
47 kΩ
1 kΩ
1 μF
(d)
300 kΩ
C
C
C
10 kΩ
R
R
R
R = 27 kΩ, C = 0.056 μF
(e)
30 kΩ
150 kΩ
820 pF
40 μH
160 μH
(f)

Figure 6.9

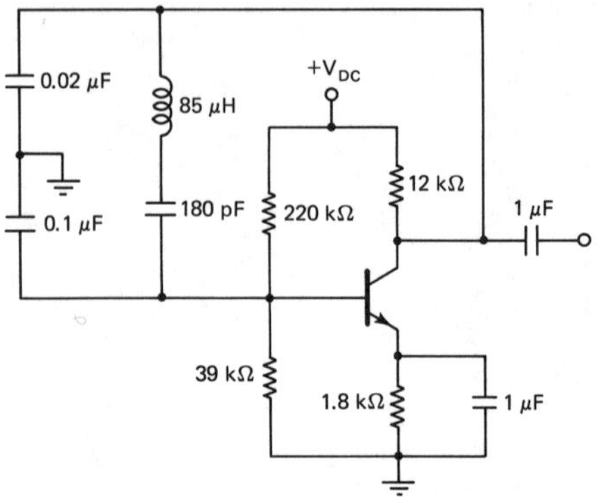

(g)

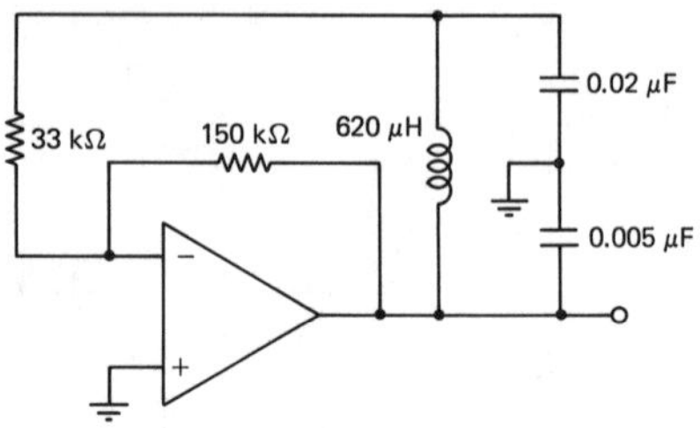

(h)

Figure 6.9 ***(Cont.)***

6.4. Identify the type of oscillator circuit and specify the frequency of resonance for each circuit shown in Figure 6.9:

	Oscillator Name	Resonant Frequency
(a)	______________	______________
(b)	______________	______________
(c)	______________	______________
(d)	______________	______________
(e)	______________	______________
(f)	______________	______________
(g)	______________	______________
(h)	______________	______________

7

Nonlinear Op Amp Applications

In addition to their many linear applications, such as amplifiers, mixers, active filters, and oscillators, op amps have proven very useful for a variety of nonlinear applications as well. In this chapter we shall investigate some op amp circuits which are mainly useful in digital applications.

7.1 THE SIMPLE COMPARATOR

In the beginning of Chapter 3, a simple gain calculation suggested that when an op amp is operated without feedback, or ''open loop,'' an input signal less than 1 mV in amplitude would normally drive the output of the op amp between maximum and minimum limits imposed by the DC supply values. This action, while undesirable for most linear uses, is quite applicable to digital situations, where ''highest'' and ''lowest'' voltages are normally the only meaningful values, used to represent binary values of zero and one.

Figure 7.1 shows a simple **comparator** circuit, in which the inverting input is grounded while the noninverting input is available for signal input. Consider the circuit's reaction to any input. Let us assume that we input a positive voltage level of 1 V to the noninverting input. With a gain of 100,000, the op amp would like to deliver an output of 100,000 volts. However, if we

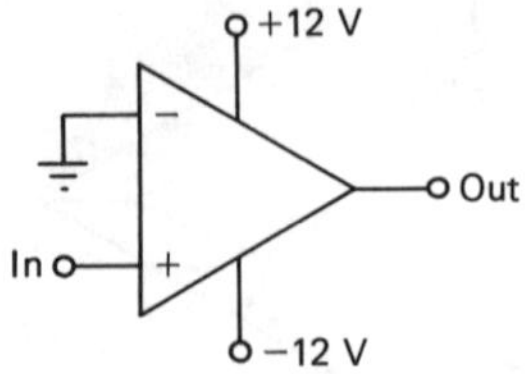

Figure 7.1

are using DC supplies of positive and negative 12 volts, the output actually cannot exceed 12 volts, and will usually go no higher (or lower on the negative side) than a maximum of 95% of the supply voltage. The output will therefore be at approximately 11 to 11.5 volts. If we increase the input voltage, no change will be seen in the output, since it is already as high as it can go, in positive saturation.

If we decrease the input amplitude, the output will remain at maximum until the input becomes small enough (less than 1 mV) to allow the output voltage to fall. So, for any input higher than 1 mV, the output will go to its positive limit. If we input a negative voltage, similar actions will occur: the output will go to its negative saturation limit with any input more negative than −1 mV. Essentially, the output will always be at maximum, when the input is higher than ground, or at minimum, when the input is lower than ground, unless our input amplitude is smaller than 1 mV. If the op amp's open-loop gain is larger, the range of inputs which do not cause saturation will be even tinier!

In Figure 7.2, we have grounded the noninverting input, and signals will be supplied to the inverting input instead. With the op amp operating open-loop, we would again expect the output to be saturated for any input amplitude of 1mV or more. This circuit's action is very much like the actions of the circuit in Figure 7.1, with one important difference. A sizable positive input will now cause negative output saturation, while a sizable negative input will now lead to positive output saturation. When using the inverting input, our output will be the opposite polarity from the input, which is a very logical action considering the inverting input's phase relationship. In this circuit, the output is low when the input is above ground, while the output goes high when the input falls below ground.

In summary, the comparator circuit generally provides a saturated output when an input of any significant size is supplied. When using the noninverting input, output polarity will match input polarity. If we instead use the inverting

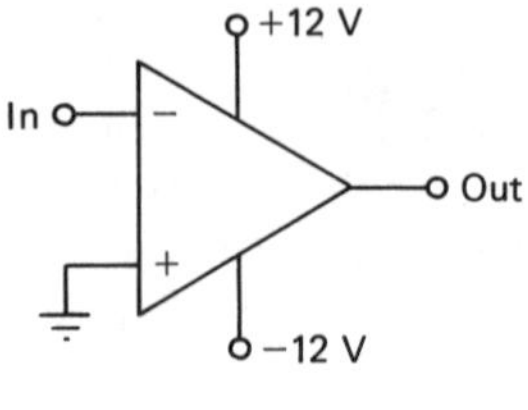

Figure 7.2

input, output polarity will be the opposite of input polarity. In the simple comparator, the unused input is always grounded.

7.2 ADJUSTABLE-THRESHOLD COMPARATORS

In the simple comparator, our output will switch states when our input crosses zero volts. The addition of a resistive voltage divider allows the selection of any voltage level (within our supply limits, of course) as the **threshold of detection.**

In Figure 7.3, a voltage divider has been added to the simple comparator circuit from Figure 7.1. If the two resistors are equal in resistance value, the inverting input will now be held at +6 V, rather than ground. An input of more than +6 V will drive the output into positive saturation, while an input of less than +6 V will drive the output into negative saturation. Even though the input may still be positive, when it is below +6 V, it is below the reference voltage level. Our threshold no longer occurs at 0 V but rather at +6 V.

By selecting appropriate resistance values, the threshold voltage, equal to the voltage across R2, may be set to any positive value within the DC source limits. If a potentiometer is used instead of the fixed resistors, the sensing threshold may be adjusted as needed. As an example, if R_1 in Figure 7.3 is 10 kΩ and R_2 is 30 kΩ, the sensing level will be at +9 volts, so that output changes will occur only when the input crosses this voltage level.

In Figure 7.4, a voltage divider has been added to the simple comparator circuit from Figure 7.2. This time, the voltage divider is connected between both DC sources, so that by appropriate resistor selections we can set our threshold voltage level at any positive or negative value within the source limits! If R_1 is 10 kΩ and R_2 is 5 kΩ, then we have a total of 24 V (12 minus-12) across

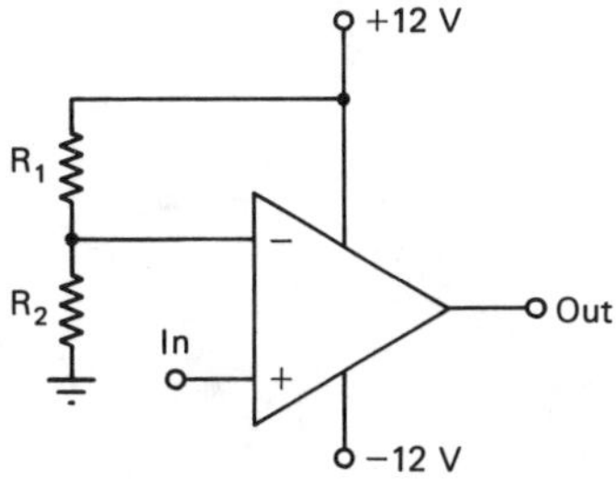

Figure 7.3

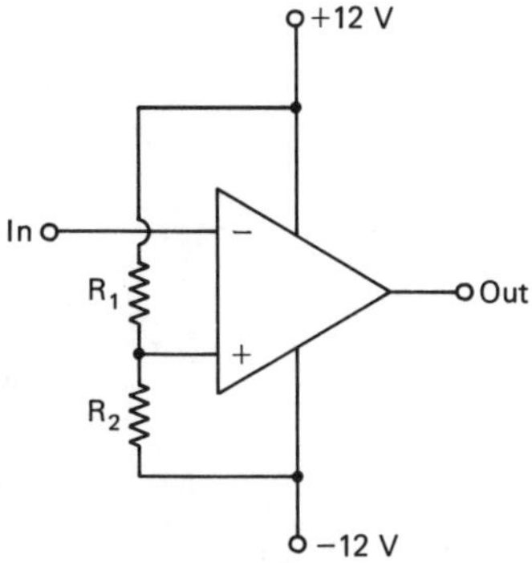

Figure 7.4

a total of 15 kΩ in our divider, for a current of 1.6 mA through each resistor. R_1 develops a 16-V drop and R_2 an 8-V drop. Therefore, with respect to ground, our sensing threshold is at −4 V. Any input above −4 V will drive the output to negative saturation, while any input below −4 V (more negative) will drive the output to positive saturation. Again, the output polarity is opposite that of the input, since we are using the inverting input.

The use of a voltage divider to set the sensing threshold level is also directly applicable to the op amp comparator operating from only one DC supply. In Figure 7.5, with supply values of +5 V and ground, the output will be either near +5 V or near 0 V. The threshold can be set at any value between 0 V and 5 V through selection of appropriate resistance values. Either input can be used, depending on the desired relationship (in or out of phase) between the input and output voltage levels.

The slew-rate characteristic of the op amp itself will always limit the speed at which the output may change to a certain maximum rate. High-speed digital circuits involving comparators will sometimes require op amps with extremely fast slew rates, in order to "keep up" with system timing needs. Some op amps are produced for high-speed digital comparator applications, with huge gain figures and extremely rapid slew rates to minimize switching delays.

One additional comparator limitation involves insuring that input signal amplitudes do not exceed the DC supply values. Inputs which exceed these limits may permanently damage the op amp. Some op amp circuits incorporate diodes between each input and the DC supply terminals. Connected to be reverse-biased for normal input voltages, these diodes become forward-biased only if the input amplitude exceeds either DC supply limit. If this occurs, the active diode will not allow the input to exceed the supply limit by more than a fraction of a volt, minimizing the possibility of damage to the op amp.

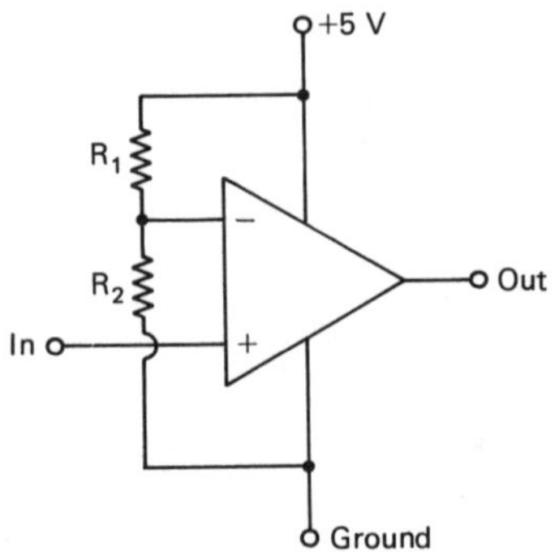

Figure 7.5

7.3 TIME-DELAY CIRCUITS

Digital circuits occasionally require timed delays, so that logic sequences are processed in a particular order. One simple circuit involves the use of a comparator to sense capacitive charge or discharge rates. Figure 7.6 shows a circuit which, when its input changes from low to high, keeps its output low until a certain time has elapsed. A low input discharges the capacitor through the diode, and keeps its voltage low. When the input goes high, the diode turns off, allowing the capacitor to begin charging. When the capacitor voltage rises beyond the threshold voltage at the inverting input, the output will go high. The amount of delay time depends on two factors: the capacitive time constant and the threshold voltage level. This type of delay circuit, when used, normally provides delays of microseconds up to milliseconds, and normally uses a capacitance value much less than 1 μF, to insure that the rapid discharge of the capacitor, which occurs each time the input goes low, does not generate too large a current flow.

Figure 7.7 shows another delay circuit which, when its input changes from high to low, delays the output from going low until after an elapsed time. In this circuit, a high input will keep the noninverting input higher than the

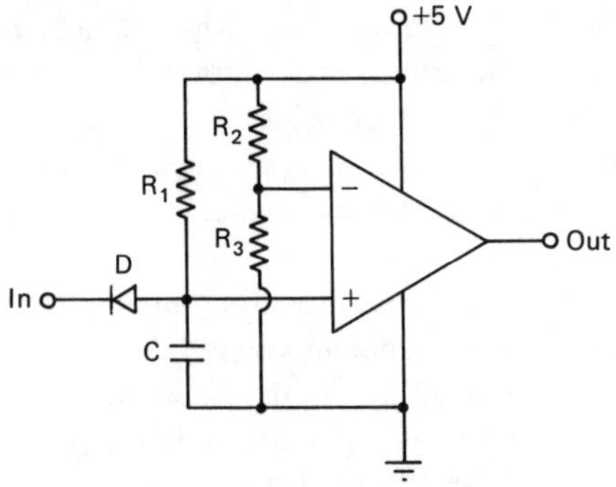

Figure 7.6

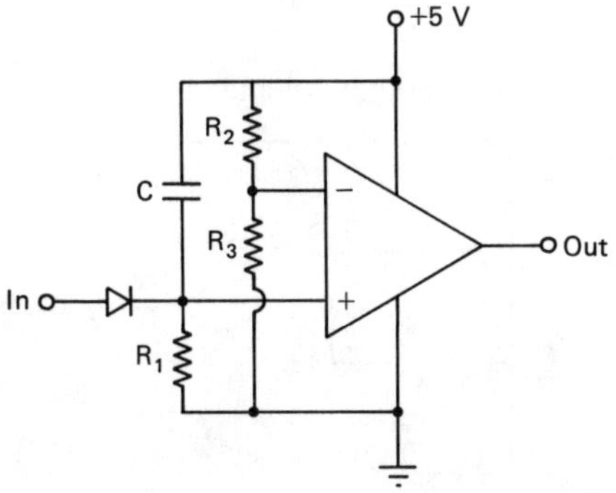

Figure 7.7

threshold voltage, causing the output to be high as well. When the input goes low, the accumulation of charge in the capacitor will cause the noninverting input voltage to fall. When it drops below the threshold voltage value at the inverting input, the ouptut will change from high to low.

Exchanging the two input connections in Figure 7.6 will give a circuit in which a low input will give a high output. When the input goes high, the output will delay in going low. Exchanging the two inputs in Figure 7.7 will provide a circuit in which a high input will give a low output. When the input goes low, the output will delay in going high. Thus, any low/high relationship between input and output can be achieved with one of these configurations.

7.4 SQUARE-WAVE OSCILLATORS

Sinusoidal oscillators are generally used in analog applications, such as audio or radio circuitry. Digital circuits, however, generally require square or rectangular oscillation waveforms. Oscillators which provide square or rectangular waves are often called **multivibrators.** Figure 7.8 shows a very simple op amp version of a circuit called the **astable multivibrator.**

Notice the presence of two separate feedback paths in this circuit. At the inverting input, the capacitor charges through R_1, while R_2 and R_3 provide a fraction of the output voltage to the noninverting input. The output of the op amp, operating open-loop, will be in either positive or negative saturation at any time. R_2 and R_3 are selected to divide the saturated output voltage level down to deliver a maximum amplitude of approximately 1 V to the noninverting input, in the same polarity as the output.

When the output goes positive, the noninverting input goes positive with it. The capacitor will also begin to accumulate voltage through R_1, taking the inverting input gradually positive with it. With the op amp working as a comparator, the output will remain in positive saturation while the capacitor voltage remains below 1 V. As discussed in Chapter 3, the voltage difference between the two inputs must fall well below 1 mV before the output may leave saturation.

When the capacitor accumulates enough voltage to take the inverting input

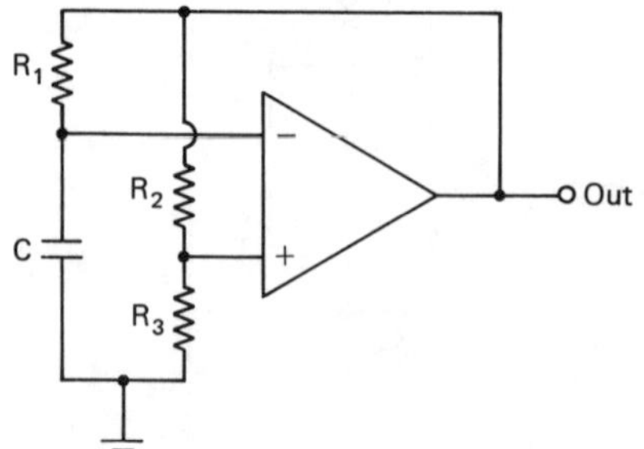

Figure 7.8

higher in voltage than the noninverting input, the output switches to negative saturation. This in turn reverses the polarity of voltage on the noninverting input. The capacitor, through R_1, will now begin charging in the negative direction. When the capacitor voltage brings the inverting input voltage level below the noninverting input voltage level, the output switches back to positive saturation. The entire cycle now repeats continuously, generating a regular, repetitive waveform.

The frequency delivered by this circuit not only depends on the values of R_1 and C, but also is affected by R_2 and R_3, which set the threshold voltage value at which output polarity reverses. With capacitive charge occurring at a rate determined by R_1 and C, a small threshold voltage level will be reached more quickly than a large threshold voltage level. Thus, an increase in the threshold level will cause an increase in the time elapsed between output reversal, and thus a decrease in frequency. A decrease in the threshold level, in decreasing the time elapsed between output reversal, serves to increase the frequency. The elapsed time that the output spends at one polarity of saturation, between switching transitions, is:

$$t = \frac{R_1 \cdot C}{1 + \dfrac{R_2}{2 \cdot R_3}} \tag{7.1}$$

Essentially, equation (7.1) defines the time which elapses between output reversals, so it is the time needed for one alternation to occur at the output. Since one full cycle contains two alternations, the period of one full wave would have to be double the elapsed time from equation (7.1):

$$P = \frac{2 \cdot R_1 \cdot C}{1 + \dfrac{R_2}{2 \cdot R_3}} \tag{7.2}$$

Frequency is simply the reciprocal of the period. The frequency of oscillation for this circuit, the reciprocal of equation (7.2), is:

$$f = \frac{1 + \dfrac{R_2}{2 \cdot R_3}}{2 \cdot R_1 \cdot C} \tag{7.3}$$

EXAMPLE 7.1

If the circuit in Figure 7.8 has a C of .1 μF, R_1 and R_3 are 10 kΩ, and R_2 is 100 kΩ, what is the oscillation frequency?

$$f = \frac{1 + \dfrac{100\text{ k}\Omega}{(2 \cdot 10\text{ k}\Omega)}}{2 \cdot 10\text{ k}\Omega \cdot .1\ \mu\text{F}} = 3000\text{ Hz}$$

In limiting the rate at which the output voltage can change, the op amp's slew-rate characteristics will force output transitions to occur gradually. While this is insignificant at low frequencies, at higher frequencies the output waveshape can depart visibly from the expected square waveform. This effect can be minimized at higher frequencies by using an op amp with a very fast slew rate.

7.5 THE INTEGRATOR

The **integrator** circuit provides an electrical action which, within its DC supply limits, approximates the mathematical process of integration. To a mathematician, integration means finding the area under a curve. Electrically, the integrator will provide a steady, unchanging output when the input is zero. When a nonzero voltage is applied to the input, the op amp will deliver a constant current to the capacitor, causing a gradual change in its voltage. This change will proceed opposite to the polarity of the input, so that a positive input causes the output to decrease, while a negative input will lead to a rising output voltage.

As long as the output does not go into saturation, the integrator responds to nonzero inputs by providing a change in output voltage. The rate of change is proportional to the size of the input voltage. Figure 7.9 shows a typical

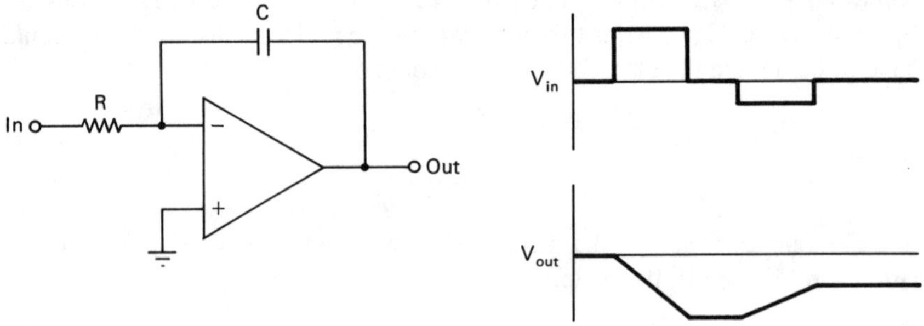

Figure 7.9

integrator circuit, accompanied by examples of typical input and output waveforms.

7.6 THE DIFFERENTIATOR

The **differentiator** circuit provides an electrical action which, within its DC supply limits, approximates the mathematical process of differentiation. To a mathematician, differentiation means finding the rate of change. Electrically, the differentiator provides an output voltage only during a change in the input

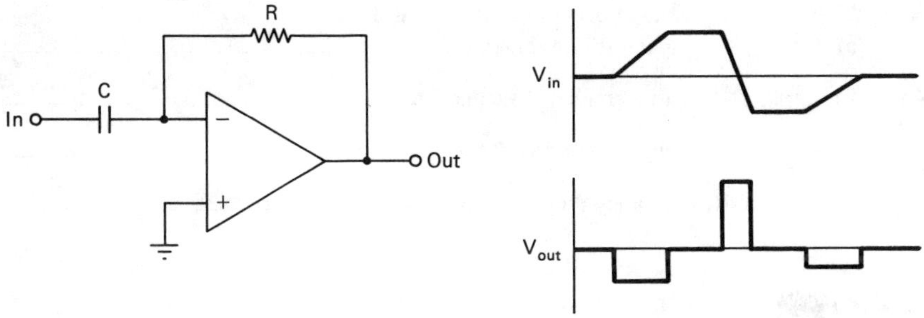

Figure 7.10

voltage. No output will appear when the input is a steady, unchanging voltage level.

As long as the output does not enter saturation, the differentiator will provide an output voltage during a change in the input voltage. The output amplitude is proportional to the rate of change at the input, so that a slow input change causes a small output, while a quick input change causes a large output. The polarity of the output will always be opposite to the direction of change, so that a rising input will cause a negative output, while a falling input will cause a positive output. Figure 7.10 shows a simple differentiator circuit, accompanied by examples of typical input and output waveforms. Note that the capacitor and resistor are in the opposite positions to those seen in Figure 7.9, the integrator.

PROBLEMS

7.1. The comparator's output voltage should always be: ______________
- **(a)** Near ground potential
- **(b)** Saturated high or low
- **(c)** Opposite to polarity of input
- **(d)** In oscillation

7.2. What type of feedback does a comparator circuit use? ______________
- **(a)** Positive feedback
- **(b)** Negative feedback
- **(c)** None: it operates open-loop

7.3. Select the correct statement about comparators: ______________
- **(a)** Output voltage change occurs when input rises or falls past the sensing voltage level.
- **(b)** Input levels should never exceed DC supply values.
- **(c)** Sensing threshold can be at any voltage level within limits of DC supply values.
- **(d)** All of the above statements are correct.
- **(e)** None of the above statements are correct.

7.4. What frequency change occurs in Figure 7.8 if we:

(a) Change the capacitor to a larger value? ______

(b) Replace R_2 with a smaller resistance value? ______

(c) Replace R_1 with a larger resistance value? ______

(d) Replace R_3 with a smaller resistance value? ______

7.5. How is the delay changed in Figure 7.6 if we:

(a) Change the capacitor to a smaller value? ______

(b) Replace R_2 with a larger resistance value? ______

(c) Replace R_1 with a smaller resistance value? ______

(d) Replace R_3 with a larger resistance value? ______

(e) Change the DC supply to a higher value? ______

7.6. Select the true statement about integrators: ______

(a) Output appears while the input is changing.
(b) Output polarity is the same as input polarity.
(c) Output rises when input is negative.
(d) Output size is proportional to rate of input change.
(e) None of these is correct.

7.7. Select the true statement about differentiators: ______

(a) Output appears while the input is nonzero.
(b) Output polarity is the same as input polarity.
(c) Output rises when input is negative.
(d) Output size is proportional to rate of input change.
(e) None of these is correct.

7.8. $R_1 = 150\ \text{k}\Omega$ and $R_2 = 30\ \text{k}\Omega$ in the circuit in Figure 7.3. What type of output level would be expected to occur for each given input level?

(a) $V_{in} = +5$ V: $V_{out} =$ ______ **(b)** $V_{in} = -6$ V: $V_{out} =$ ______

(c) $V_{in} = -1$ V: $V_{out} =$ ______ **(d)** $V_{in} = -9$ V: $V_{out} =$ ______

(e) $V_{in} = 0$ V: $V_{out} =$ ______ **(f)** $V_{in} = +2$ V: $V_{out} =$ ______

7.9. Referring to the circuit in Figure 7.8, determine the frequency of oscillation for each set of component values:

	R_1	R_2	R_3	C	Frequency
(a)	12 kΩ	33 kΩ	56 kΩ	470 pF	________
(b)	820 Ω	2.7 kΩ	4.7 kΩ	15 μF	________
(c)	75 kΩ	180 kΩ	120 kΩ	82 pF	________
(d)	6.8 kΩ	3.9 kΩ	1.5 kΩ	2.2 μF	________
(e)	910 Ω	820 Ω	510 Ω	.15 μF	________
(f)	9.5 kΩ	24 kΩ	6.8 kΩ	33 pF	________

8

Op Amp Voltage Regulators

The huge-gain, large input impedance, and small output impedance of the op amp make it useful not only for applications in which output variation occurs, but also for those in which a constant output level is essential. In voltage-regulator circuits, where the output voltage amplitude should be as constant as possible, the op amp can be used as a voltage regulator in a variety of interesting ways. The output levels of these regulators must always be smaller than the DC supply voltage levels when the op amp supplies the load current directly. In other regulators, the op amp only provides error detection to control load current and voltage through a transistor.

8.1 STANDARD REFERENCE VOLTAGES

Comparison of an input voltage level to a known **reference level** is used in some digital voltmeters, for example, to develop the digital display of DC voltage measurements. The reference voltage level, provided by a special battery, must be kept as precise and constant as possible. If the reference level varies, the accuracy of the meter will suffer.

Figure 8.1 shows a unity-gain op amp with the standard reference battery connected between ground and the noninverting input. With its large input

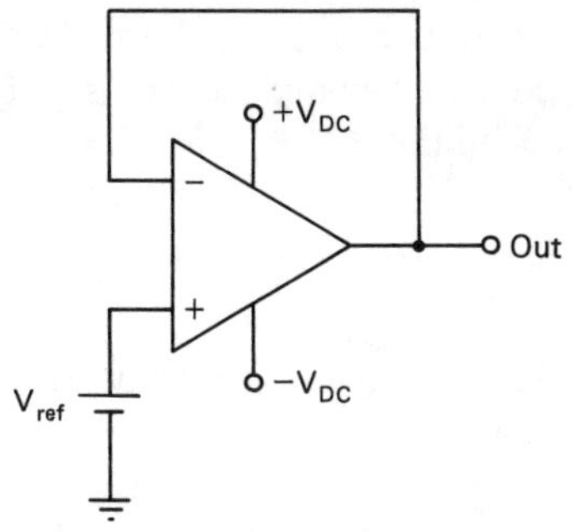

Figure 8.1

impedance, the op amp will draw only a very tiny current from the battery. The op amp's output, in the unity-gain format, will provide an output voltage *equal* to the reference source. In addition, the op amp can deliver any output current level within its maximum ratings, while drawing virtually no current from the reference source. The op amp has effectively isolated the reference source from any large current requirements, which serves to minimize operational voltage changes, and also prolongs the useful life of the reference source. By reversing the polarity of the reference cell, a negative regulated voltage can also be provided by this circuit.

Figure 8.2 shows a circuit which, like Figure 8.1, also provides isolation of its reference source from large current demands in exactly the same way. However, the presence of the resistive feedback network in this circuit suggests that the op amp is connected in the noninverting amplifier format. This conclusion is quite correct! This circuit, in addition to isolating the reference source from any current requirements, is also arranged to provide a voltage gain. As a result, this current delivers a DC output level which is *larger* than the reference voltage, defined by:

$$V_{\text{out}} = V_{\text{ref}} \cdot \left(1 + \frac{R_f}{R_i}\right) \tag{8.1}$$

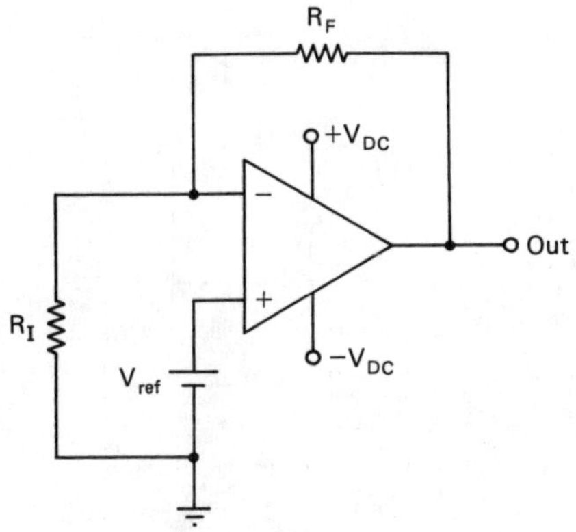

Figure 8.2

By proper resistor selection, a small reference level can be multiplied to any larger output voltage value, as long as it falls within the DC supply limits as well as within the op amp's output capabilities.

EXAMPLE 8.1

R_f is 10 kΩ and R_i is 2 kΩ in Figure 8.2. If we connect a 1.5-V battery to the noninverting input, what is the output voltage?

$$V_{out} = 1.5 \cdot \left(1 + \frac{10{,}000}{2000}\right) = +9\text{ V}$$

8.2 OP AMP ZENER REGULATORS

Instead of a standard reference voltage source, a **zener diode** may be used to set the reference voltage level for the op amp, as shown in Figure 8.3. The zener must conduct a small reverse current at all times if it is to provide its highly predictable zener voltage to the op amp. This small current is provided by the resistor in series with the zener. The unity-gain op amp format, with the zener voltage value at its input, will provide the same voltage at the output, and, like the format in Figure 8.1, it also isolates the zener from load current variations. When used within the op amp's maximum output current limits, this regulator circuit will provide an extremely stable output voltage.

The output in Figure 8.3 will be a positive voltage. In Figure 8.4, the zener and its series resistor have instead been connected between ground and the negative DC source. In placing negative voltage at the noninverting input,

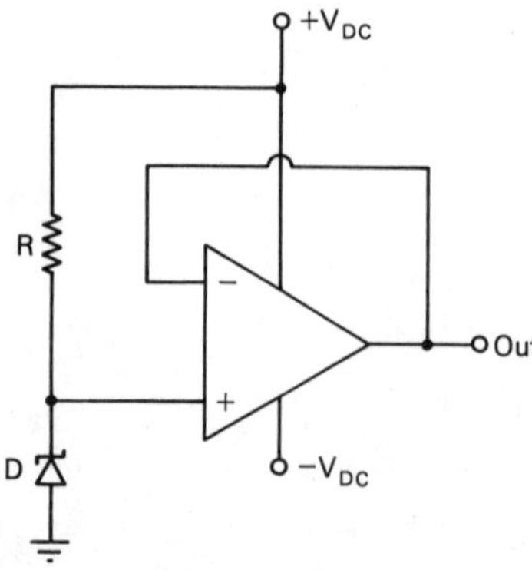

Figure 8.3

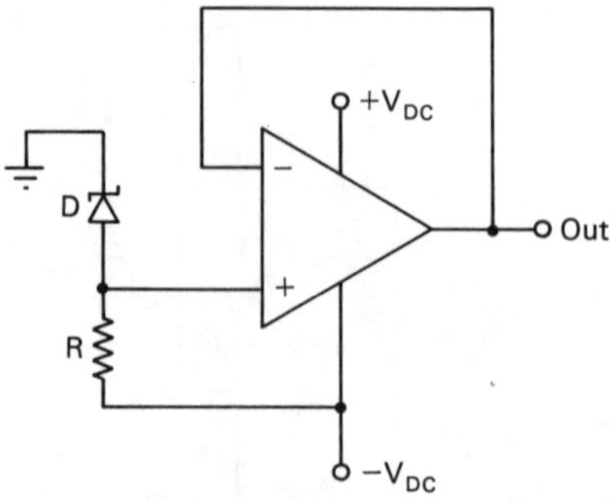

Figure 8.4

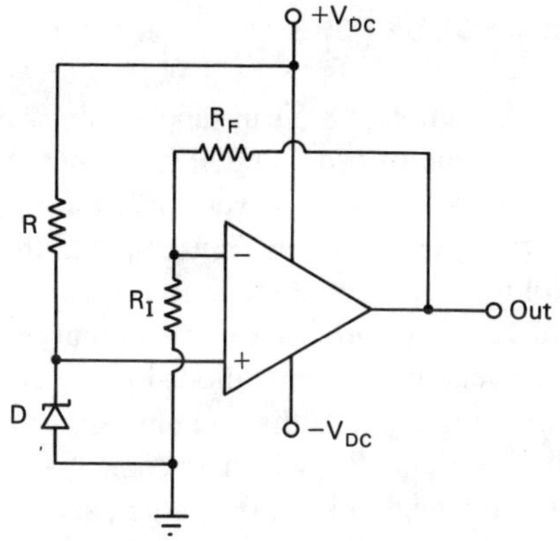

Figure 8.5

a negative output voltage will be provided rather than a positive output voltage. Except for output polarity, this circuit will otherwise function just like the circuit in Figure 8.3.

Figure 8.5 shows the circuit from Figure 8.3 modified by the addition of the standard resistive feedback network. Through the same actions as those seen with Figure 8.2, as long as the output remains below saturation levels, the output voltage will be determined by:

$$V_{out} = V_{zener} \cdot \left(1 + \frac{R_f}{R_i}\right) \tag{8.2}$$

EXAMPLE 8.2

Refer to Figure 8.5 for these questions:

(a) If R_f is 33 kΩ, R_i is 22 kΩ, and the zener rating is 3.2 V, what is the output?

$$V = 3.2 \cdot \left(1 + \frac{33{,}000}{22{,}000}\right) = 3.2 \cdot 2.5 = 8 \text{ V}$$

(b) If R_f is 5.6 kΩ, R_i is 2.0 kΩ, and the zener rating is 1.5 V, what is the output?

$$V = 1.5 \cdot \left(1 + \frac{5600}{2000}\right) = 1.5 \cdot 3.8 = 5.7 \text{ V}$$

By adding the same feedback network to the circuit in Figure 8.4, the negative input voltage level will provide a negative output voltage which is also $(1 + R_f/R_i)$ times larger than the input. As in Figure 8.2, the input zener voltage can be multiplied to any larger output voltage within the op amp's output limits, through appropriate resistor value selection.

8.3 OP AMPS IN SERIES REGULATORS

The regulators shown in Figures 8.1 through 8.5 can supply only as much output current as the op amp itself is capable of providing at its output. When larger output current capabilities are required, a large power transistor replaces the op amp as the source of output current, as shown in Figure 8.6. In this event, the op amp, serving only to control the transistor, functions as a comparator rather than an amplifier, in order to detect any variations in the output voltage level.

This regulator, like most others, uses a zener diode to set a constant reference voltage level at the noninverting input of the op amp. Reverse current for the zener is constantly supplied through R_1, which is connected to the input voltage, R_2 and R_3 form a series voltage divider at the output, serving to provide a specific fraction of the output voltage to the inverting op amp input. The NPN transistor is connected with its collector to the input voltage terminal, its emitter to the output voltage terminal and its base to the output of the op amp. It must provide not only the bleeder current for the R_2/R_3 voltage divider but also supply all required load current.

This transistor will usually be mounted on a heatsink, particularly when the output current capability is 1 A or more. With a voltage drop across it equal to the difference between the input and output voltage levels, a large load current flow will result in a large power dissipation within the transistor. For example, if the input voltage is 18 V, the output voltage is 10 V, and a load current of 4 A flows, the voltage drop across the transistor will be 18 minus 10, or 8 V. The transistor, with 8 V across it and 4 A through it, will dissipate 32 watts!

When input voltage appears, R_1 supplies a reverse current through the zener, which will provide its rated reverse voltage level to the noninverting op amp input. Responding as a comparator, the op amp output will be driven high, providing base drive to the transistor. As output current flows, the voltage from the R_2/R_3 divider will rise, raising the inverting input voltage level as well. When this divider voltage level reaches the zener voltage, the op amp's output will drop from maximum saturation, since both input voltages are now at the

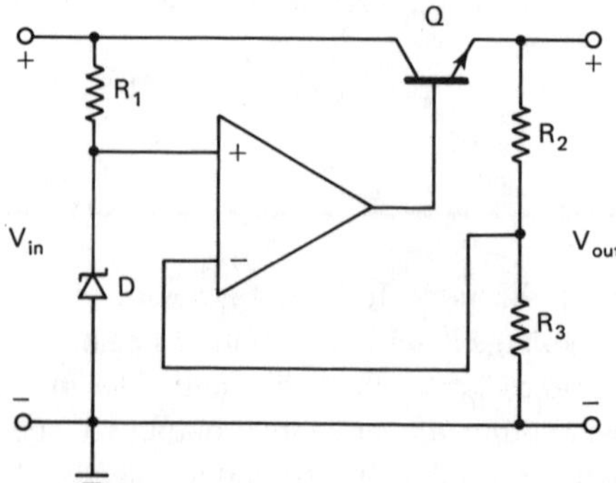

Figure 8.6

same level, which decreases the transistor's base drive from the initial maximum during turn-on to its usual operating level.

If the output voltage becomes a little too large, the inverting input will be raised above the zener voltage, causing the op amp's output to decrease in voltage. With decreased base drive, the transistor will decrease the output current supplied, forcing the output voltage back downward to the normal value. If the output voltage falls, the voltage at the inverting input will also fall below the zener voltage, driving the op amp's output higher in voltage. With increased base drive, the transistor will provide more current, forcing the output voltage back upward to the normal value.

In this regulator, the zener diode, while an essential part of the voltage regulation system, does not directly determine the output voltage. In fact, the output voltage, while it must be smaller than the input voltage, will usually be larger than the zener rating, V_z, according to this simple relationship:

$$V_{\text{out}} = V_z \cdot \left(1 + \frac{R_2}{R_3}\right) \tag{8.3}$$

EXAMPLE 8.3

Refer to Figure 8.6 for these questions:

(a) If R_2 and R_3 are 22 kΩ and the zener rating is 6 V, the output voltage will be:

$$V = 6 \cdot \left(1 + \frac{22{,}000}{22{,}000}\right) = 6 \cdot 2 = 12 \text{ V}$$

(b) If R_2 is 1 kΩ, R_3 is 2 kΩ, and the zener rating is 10 V, the output voltage will be:

$$V = 10 \cdot \left(1 + \frac{1000}{2000}\right) = 10 \cdot 1.5 = 15 \text{ V}$$

(c) If R_2 is 18 kΩ, R_3 is 6 kΩ, and the zener rating is 1.5 V, the output voltage will be:

$$V = 1.5 \cdot \left(1 + \frac{18{,}000}{6000}\right) = 1.5 \cdot 4 = 6 \text{ V}$$

The zener rating is normally selected to fall between one-third and two-thirds of the input voltage size, insuring that R_1 will always have sufficient voltage across it to provide the constant flow of reverse current through the zener. The output voltage, as the absolute minimum, may be made equal to the zener voltage, in which case the output voltage divider can be deleted, since

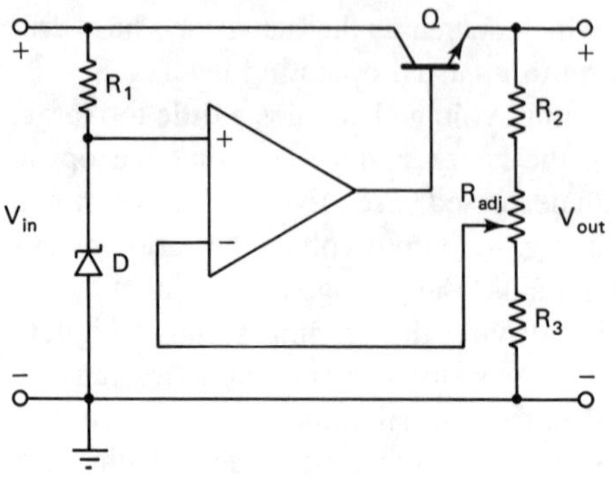

Figure 8.7

the inverting op amp input is in this case directly connected to the positive output terminal.

Some regulators contain a potentiometer which is connected, as shown in Figure 8.7, with its fixed element placed in series between R_2 and R_3, while the wiper is connected to the inverting op amp input. The op amp drives the transistor in such a way to keep both op amp inputs at the same voltage level. This fact suggests that movement of the pot's wiper, which changes the fraction of the output voltage developed at the inverting input, will force a change in the output voltage such that it brings both input voltage levels back together. If the wiper is moved toward R_2, a larger fraction of the output voltage will appear at the inverting input, forcing the output voltage to decrease in compensation. Movement of the wiper toward R_3, providing a smaller fraction of the output voltage to appear at the inverting input, forces the output voltage to rise in compensation.

In Figure 8.8, two additional components, Q2 and R_5, have been added to the circuit. These additional components enable the output current to be

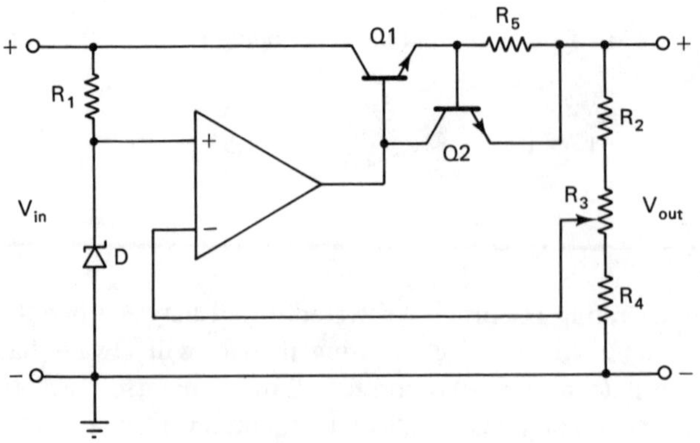

Figure 8.8

limited to a maximum safe value. Note that R_5, a small resistance in series with Q1's emitter, also appears in parallel with Q2's base-emitter junction.

When load current flows, it must flow through R_5, creating a voltage drop across it. Control of current-limiting action in this circuit arises directly from the presence of R_5 in parallel with Q2's base-emitter junction. The voltage across R_5, proportional to the load current drawn, remains smaller than Q2's barrier potential when the load current is less than the maximum amount, so Q2 does not conduct, and Q1 will receive all available base current.

If R_5's voltage drop becomes sufficiently large, approximately .7 V, it will forward-bias Q2's base-emitter junction enough to turn Q2 on. Q2 will then conduct the excess portion of Q1's base drive through its collector path. Any further increase in load current will increase the voltage across R_5 even more, turning on Q2 even more. Q2 will then draw even more current away from Q1's base. Q1's base drive and resulting load current will thus be limited to a specific maximum amount. Q2 will turn off only when the load current through R_5 decreases enough to allow the voltage drop across R_5 to fall below Q2's barrier potential.

The resistance value of R_5 is a critical selection in this circuit, since it alone will determine the maximum load-current figure at which current limiting will occur. Since Q2 can provide limiting action only when R_5's voltage rises to .7 V and turns it on, the selection of the resistance value of R_5 determines the current level at which its voltage drop approaches .7 V. Selecting the resistance of R_5 to provide a specific maximum current-limiting level is actually quite an easy matter, since under limiting conditions its voltage drop must be equal to Q2's base-emitter voltage drop of approximately .7 V. The resistance value must then be:

$$R_5 = \frac{.7\ \text{V}}{I_{\text{max}}} \tag{8.4}$$

EXAMPLE 8.4

We wish to construct the regulator circuit in Figure 8.8, with current limiting to occur at a load current of 5 A. What would R_5 need to be?

$$R_5 = \frac{.7\ \text{V}}{5\ \text{A}} = .14\ \Omega$$

When the value of R_5 is known, the maximum current level at which limiting occurs is also quite easy to find:

$$I_{\text{max}} = \frac{.7\ \text{V}}{R_5} \tag{8.5}$$

EXAMPLE 8.5

Refer to the regulator in Figure 8.8. What maximum load current will be available if:

(a) R_5 is 1.4 Ω? I_{max} will be: ______________________

$$I_{max} = \frac{.7\ \text{V}}{1.4\ \Omega} = \tfrac{1}{2}\ \text{A or } 500\ \text{mA}$$

(b) R_5 is .1 Ω? I_{max} will be: ______________________

$$I_{max} = \frac{.7\ \text{V}}{.1\ \Omega} = 7\ \text{A}$$

The larger the desired current-limiting figure, the smaller the resistance value of R_5 must be, so that its voltage drop stays below .7 V with less than maximum current levels.

8.4 OP AMPS IN SHUNT REGULATORS

In series regulators, load voltage is controlled by an active device placed in series with the load, so that the active device must carry the entire load current flow and withstand the resulting power dissipation. In a **shunt regulator,** the active device will instead be found in parallel with the load, so that it can shunt current around the load when it is necessary. Shunt regulators are usually used for applications which require large but reasonably constant load currents.

Figure 8.9 shows a typical shunt regulator circuit which contains an op amp. The function of the op amp is that of error detection, not unlike its function in the series regulators. A zener diode, kept in reverse conduction by R_1, maintains a constant voltage at the op amp's inverting input, to which it is connected. A voltage divider, consisting of R_3 and R_4, provides a specific fraction of the output voltage level to the noninverting input. The op amp's output is connected to the base of the shunt transistor, which has its collector and emitter terminals connected directly in parallel with the load.

An additional resistor appears in series between the input and output terminals, in the position that was occupied by the transistor in the series regulators. In the shunt regulator, this series resistor will withstand the voltage drop between the input voltage level and the output voltage level, and it must also provide a constant current level which supplies both the load and shunt currents. This implies that in this circuit, R_2 will probably dissipate a large amount of power, so it must be rated to handle this large dissipation. The shunt transistor, however, will normally carry current levels which are much smaller than the

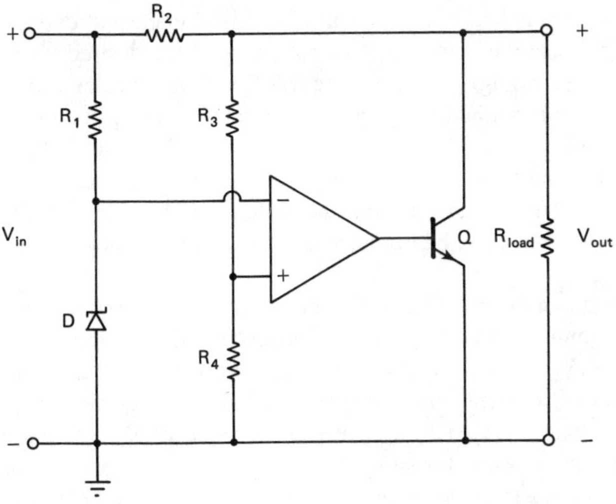

Figure 8.9

load current, so the shunt transistor normally dissipates much smaller amounts of power than the series resistor must handle.

When an input voltage is supplied to the circuit, R_1 will provide the zener with reverse current flow, so the zener will maintain its constant zener voltage rating at the op amp's inverting input. Ideally, the output voltage level will provide a voltage at the op amp's noninverting terminal which exceeds the zener voltage level just enough to drive the op amp's output to a positive level which is sufficient to turn on the shunt transistor just slightly. The actual load voltage level provided will depend on the zener voltage rating and the particular fraction of the load voltage provided to the op amp by the R_3/R_4 voltage divider, according to this formula:

$$V_{\text{out}} = V_{\text{zener}} \cdot \left(1 + \frac{R_3}{R_4}\right) \qquad (8.6)$$

The series resistor R_2 provides a constant current which is somewhat larger than the load's normal current requirement. This constant current represents the sum of load current and shunt current at all times. Voltage regulation occurs in this circuit through delivery of just enough current through the load to cause the correct voltage level. The transistor must then shunt the remainder of the total current from R_2 around the load.

If the load voltage increases, the R_3/R_4 voltage divider will also raise the voltage at the noninverting op amp terminal, which will cause the op amp's output voltage to rise as well. Since this will increase the base drive to the shunt transistor, the transistor will draw a larger shunt current. Since the sum of shunt and load currents is constant, this will force the load current to decrease, which serves to decrease the output voltage back toward the desired level.

If the load voltage decreases, the R_3/R_4 voltage divider will also decrease the voltage at the noninverting op amp terminal, causing the op amp's output voltage to fall as well. Since this will decrease the base drive to the shunt transistor, the transistor will draw a smaller shunt current. Since the sum of shunt and load currents is constant, this will allow the load current to increase, which serves to increase the output voltage, bringing it back toward the desired level.

By virtue of the presence of the series resistor R_2, through which all current must pass, shunt regulators provide a current-limiting action controlled by the resistance of R_2. If the load should become shorted, R_2's series resistance will control the total current flow, which will limit at a maximum value equalling the input voltage divided by the resistance value of R_2. In this instance, all current will pass through the shorted load, leaving no current flow and therefore no power dissipation in the shunt transistor. R_2, however, with the entire input voltage across it, dissipates more power in this situation than in any other.

The worst situation for a shunt regulator occurs when the load is unconnected, or open. Since no current flows through the open load, this forces all the current to pass through the shunt transistor, causing maximum power dissipation to occur within the transistor in this situation. Shunt regulators, by virtue of their construction, are designed to supply load current at all times. Operating a shunt regulator without its load connected places extreme stress on the shunt transistor and can easily damage it permanently.

EXAMPLE 8.6

In a shunt regulator circuit as shown in Figure 8.9, we have the following component values:

$$\text{zener diode rating} = 4\text{ V}, \qquad R_1 = R_4 = 2\text{ k}\Omega$$

$$R_2 = 1.2\ \Omega, \qquad R_3 = 1\text{ k}\Omega, \quad V_{in} = 9\text{ V}$$

(a) Determine the expected load voltage level:

$$V_{out} = 4\text{ V} \cdot \left(1 + \frac{1\text{ k}\Omega}{2\text{ k}\Omega}\right) = 4\text{ V} \cdot 1.5 = 6\text{ V}$$

(b) Determine the maximum load current available:

I_{max} depends on R_2's voltage and resistance values. With V_{in} at 9 V and V_{out} at 6 V, R_2 has 3 V across it:

$$I_{max} = \frac{3\text{ V}}{1.2\ \Omega} = 2.5\text{ A}$$

(c) How much power does the zener diode dissipate?

Current through the zener is the current through R_1. If the zener conducts, R_1 will have 5 V across it, with $I = 5\ \text{V}/2\ \text{k}\Omega$ or 2.5 mA. The diode's power dissipation is then:

$$P_{zener} = 4\ \text{V} \cdot 2.5\ \text{mA} = 10\ \text{mW}$$

(d) If the load should become shorted, how much power will R_2 be forced to dissipate?

If the load shorts, R_2 receives the entire input of 9 V across it. With 9 V across 1.2 Ω, 7.5 A will flow. Thus, R_2 will dissipate:

$$P_{R_2} = 9\ \text{V} \cdot 7.5\ \text{A} = 67.5\ \text{W}$$

(e) If the load should become open, how much power will the shunt transistor be forced to dissipate?

An open load will force all available current to pass through the transistor, so that it must dissipate:

$$P_Q = 6\ \text{V} \cdot 2.5\ \text{A} = 15\ \text{W}$$

PROBLEMS

8.1. For each circuit in Figure 8.10, specify the regulator type and the output voltage that it will provide.

	Regulator Type	Output Voltage
(a)	____________	____________
(b)	____________	____________
(c)	____________	____________
(d)	____________	____________
(e)	____________	____________

8.2. In Figure 8.6, $R_1 = R_2 = 10\ \text{k}\Omega$, $R_3 = 15\ \text{k}\Omega$, V_{in} is 15 V, and the zener is rated at 6 V:

(a) What output voltage will be provided? ____________

(b) When passing 2 A, Q will dissipate: ____________

(c) What is the current through the zener? ____________

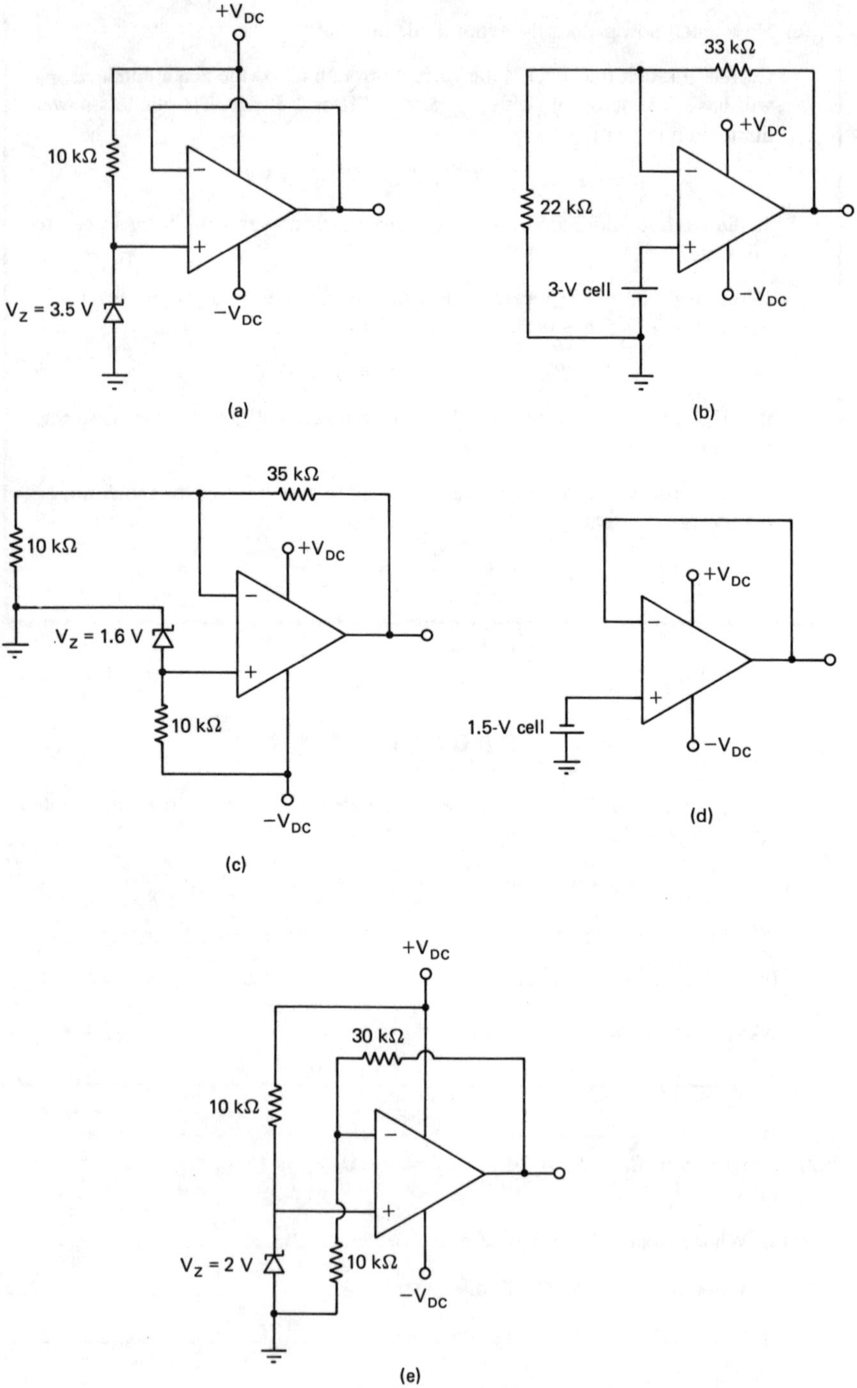

+V_DC
10 kΩ
V_Z = 3.5 V
−V_DC
(a)
33 kΩ
+V_DC
22 kΩ
3-V cell
−V_DC
(b)
35 kΩ
10 kΩ
+V_DC
V_Z = 1.6 V
10 kΩ
−V_DC
(c)
+V_DC
1.5-V cell
−V_DC
(d)
+V_DC
30 kΩ
10 kΩ
V_Z = 2 V
10 kΩ
−V_DC
(e)

Figure 8.10

(d) What amount of power does R_3 dissipate? ______

(e) If R_1 opens, what will V_{out} do? ______

(f) If R_2 opens, what will V_{out} do? ______

(g) If R_3 opens, what will V_{out} do? ______

8.3. In Figure 8.8, $R_1 = R_2 = R_4 = 10\ k\Omega$, $R_3 = 30\ k\Omega$, $R_5 = .05\ \Omega$, $V_{in} = 35$ V, and the zener rating is 5 V:

(a) Output current cannot exceed what value? ______

(b) Output voltage can be set to what maximum? ______

With maximum output voltage and current, how much power will be dissipated by Q1? ______

(c) Output voltage can be set to what minimum? ______

With minimum output voltage and maximum load current, Q1 dissipates what power? ______

(d) How much power is dissipated by the zener? ______

(e) When it limits current, R_5 will dissipate: ______

8.4. In Figure 8.9, $R_1 = R_3 = R_4 = 10\ k\Omega$, $R_2 = 1\ \Omega$, $V_{in} = 15$ V, and the zener is rated at 5 V:

(a) What voltage does the load receive? ______

(b) What total current does R_2 make available? ______

(c) How much power does the zener dissipate? ______

(d) If the load is shorted, R_2 will dissipate: ______

(e) If the load goes open, what power level will the shunt transistor be forced to dissipate? ______

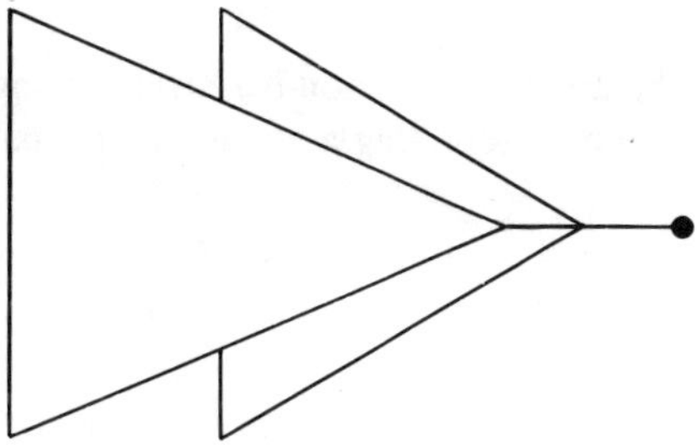

Glossary

Active device: An electronic component with nonlinear characteristics, such as a diode, transistor, or op amp. Many active devices can provide amplification.

Active filter: A circuit including one or more active devices, which exhibits frequency-dependent response.

Band-pass filter: A circuit which passes only a certain band of frequencies and rejects all others.

Band-reject filter: A circuit which rejects a certain band of frequencies and passes all others.

Bandwidth: The range of frequencies in which the gain of an amplifier is at or near the maximum value.

Bias: The application of steady DC voltage and current conditions which are essential for most active devices such as transistors.

BJT: Abbreviation for Bipolar Junction Transistor.

Buffer: An amplifier which can provide a large output current from a tiny input current, effectively isolating the input signal source from any large current demands; since its voltage gain is 1, it is also called a *unity-gain amplifier*.

Cascading: Connecting several circuits with one output going to the next input, so that the individual gains multiply together for a large overall gain.

Cascode circuit: A common-collector BJT driving a common-base BJT, giving as much overall gain as a common-emitter, but presenting a much larger input impedance.

Center frequency: The frequency associated with resonant filters. Gives maximum response in a band-pass filter, or minimum response in a band-reject filter.

Clapp oscillator: Op amp or transistor oscillator based on series *LC* resonance to set its frequency of oscillation.

Closed-loop: The presence of negative feedback in an op amp circuit which reduces gain while also expanding its bandwidth, or frequency response.

Colpitts oscillator: Op amp or transistor oscillator with two series capacitors placed in a parallel *LC* resonant circuit to determine the oscillation frequency.

Common-mode operation: An op amp receives two identical input signals, in phase with each other, providing minimal output response.

Common-mode rejection ratio: The ratio of an op amp's large differential gain to its small common-mode gain; also a measure of op amp's ability to reject common-mode inputs.

Comparator: Op amp circuit which compares an input voltage to a reference level. Saturated output indicates if input is ''higher'' or ''lower'' than the reference level.

Constant-current circuit: A circuit which always draws the same amount of current. Minimizes op amp common-mode response when used with a differential amplifier circuit.

Crystal oscillator: Op amp or transistor circuit based on mechanical resonance of a quartz crystal to set the resonant frequency.

Darlington pair: Two BJT's connected with collectors tied together and the first emitter to the second base, providing a gain figure which is the product of the two individual values.

Difference amp: Op amp format providing an output which is proportional to the difference between the two input signals.

Differential amplifier: Input stage of an op amp which responds very strongly to differential inputs but very weakly to common-mode inputs.

Differential inputs: Two simultaneous op amp input signals, one going to each input, which are out of phase with each other, ideally by 180 degrees. Normally they are of the same frequency and amplitude.

Differentiator: An op amp circuit which provides an output related to the rate of change at the input.

Direct coupling: Connection of the output of one stage directly to the next input without the need to block DC levels.

Error detector: Op amp application in series and shunt regulators in which output voltage is compared to a constant reference voltage level.

Feedback: Circuit path which delivers a portion of the output signal back to the input point; see *Positive feedback* and *Negative feedback*.

FET: Abbreviation for Field-Effect Transistor.

Frequency response: The range of frequencies over which an amplifier provides its maximum gain, also called its bandwidth.

Gain-bandwidth product: Op amp specification that is always a constant figure, such that a reduction in gain leads to a larger bandwidth, while an increase in gain leads to a reduced bandwidth, with the product of gain times bandwidth always remaining the same.

Half-power (3-dB) frequency: Critical frequency for nonresonant filters at which the output power level is reduced to half of the maximum; also called *rolloff frequency*.

Hartley oscillator: Op amp or transistor oscillator based on two series inductors placed with a capacitor in a parallel resonant circuit to detemine the oscillation frequency.

High-frequency rolloff: Falling of amplifier gain significantly below the maximum gain value above a certain high frequency, which is also the top end of the bandwidth.

High-pass filter: Filter circuit which passes high frequencies but blocks low frequencies at its output.

Input offset: A small difference in voltage placed on differential amp inputs to reduce the voltage appearing at the output, caused by internal differences between the two transistors, to zero.

Integrator: Op amp circuit in which the output changes at a rate proportional to the size of the input; an input of zero results in a steady, unchanging output.

Inverting amp: Op amp format in which input and output signals are 180 degrees out of phase; gain is determined by resistor values used in the feedback path.

Low-frequency rolloff: Falling of amplifier gain significantly below the maximum gain value below a certain low frequency, which is also the low end of the bandwidth.

Low-pass filter: Circuit which passes low frequencies but blocks high frequencies at the output.

Mixer amp: Op amp format in which two or more inputs are combined or "mixed" to a single output signal; all input resistors are the same value, so that all inputs receive the same gain.

Negative feedback: Feedback path which places a portion of the output signal back to the input, out of phase with the original input signal; serves to reduce gain.

Noninverting amp: Op amp format in which input and output signals are in phase with each other; gain is set by resistor values used in the feedback path.

Open-loop gain: The characteristic gain of the op amp without any feedback path; normally a huge figure.

Open-loop operation: With no feedback present, the op amp provides its full gain to any input; the output will usually be driven into saturation very easily.

Oscillator: A circuit which, while supplied only with a DC voltage source, internally generates and maintains an AC output signal, normally with predictable amplitude and frequency.

Output saturation: Condition where excessive input size attempts to drive the op amp's output beyond its DC supply limits; instead, the output limits at its maximum or minimum level and the waveform is lost. Such output conditions are typical with the comparator.

Passive device: An electronic component which exhibits linear relationships between voltage and current, such as a resistor or a capacitor.

Phase-shift oscillator: Op amp or transistor oscillator based on achieving a 180-degree phase shift through three cascaded *RC* networks to determine the oscillation frequency.

Piezoelectric: A material which physically deforms in response to an applied voltage and also generates a voltage in response to deformation; physical resonance determines its electrical resonant frequency.

Positive feedback: A feedback path which places a portion of the output signal back to the input in phase with it; serves to increase gain and oscillation tendencies.

Resonance: Sinusoidal interaction between an inductor and a capacitor at the particular frequency where both reactance values are exactly equal.

Rolloff: Significant decrease in amplifier gain, either above or below the critical frequencies which determine the limits of the bandwidth.

Sensing threshold: Constant DC level which is placed at one input in a comparator; the output will reverse states when the input signal at the other input rises or falls past this reference value.

Series regulator: A voltage-regulator circuit in which output voltage is supplied through a transistor which, placed in series with the load, must also conduct all output current as well.

Shunt regulator: A voltage-regulator circuit in which a large current is always available through a series resistor; a transistor placed in parallel with the load achieves voltage regulation by shunting all current in excess of load requirements around the load.

Single-ended input: Op amp has only one input signal connected to either input, with the unused input connected to ground.

Slew rate: The fastest possible rate at which the output of an op amp can change from one voltage level to another.

Summing amp: An inverting op amp format in which two or more inputs with equal gains will be combined or added to provide one single output signal; this circuit is also called the "mixer."

Unity-gain circuit: Op amp circuit in which its output is directly connected to the inverting input, giving a voltage gain of 1; also called the *buffer amp*.

Voltage regulator: A circuit which maintains a constant output voltage level at all times, even if the input voltage level or output current requirements vary.

Weighted summing amp: An inverting op amp format with two or more inputs which, having different input resistance values, provides a different gain figure for different input points; as in the mixer, all inputs will still be combined in a single output signal.

Wein-bridge oscillator: Op amp oscillator circuit which is based on the presence of a special "lead-lag" *RC* network for determination of its oscillation frequency.

Zener diode: A special-purpose diode which, when conducting under reverse bias, provides a very steady reverse voltage level for use in voltage-regulator circuits.

Bibliography

Bogart, Theodore F., *Electronic Devices and Circuits*, pp. 501–642. Columbus, Ohio: Merrill Publishing Co., 1986.

Boylestad, Robert, and Louis Nashelsky, *Electronic Devices and Circuit Theory*, 4th ed., pp. 471–562. Englewood Cliffs, N.J.: Prentice-Hall, Inc., 1987.

Brant, Carroll A., *Electronics for Communication*, pp. 298–325. Chicago, Ill.: Science Research Associates, Inc., 1983.

Cox, James F., and S. Roser Everett, *Electronic Principles—Integrated and Discrete*, pp. 196–268. Englewood Cliffs, N.J.: Prentice-Hall, Inc., 1987.

Floyd, Thomas L., *Electronic Devices*, 2nd ed., pp. 520–734. Columbus, Ohio: Merrill Publishing Co., 1988.

Hufault, John R., *Op Amp Network Design*. New York, N.Y.: John Wiley & Sons, Inc., 1986.

Maloney, Timothy, J., *Industrial Solid-State Electronics—Devices and Systems*, 2nd ed., pp. 261–285. Englewood Cliffs, N.J.: Prentice-Hall, Inc., 1986.

Metzger, Daniel L., *Electronic Circuit Behavior*, 2nd ed., pp. 246–288. Englewood Cliffs, N.J.: Prentice-Hall, Inc., 1983.

Schuler, Charles A., *Electronic Principles and Applications*, pp. 135–153. New York, N.Y.: McGraw-Hill Book Company, 1979.

Wedlock, Bruce D., and James K. Roberge, *Electronic Components and Measurements*, pp. 240–258. Englewood Cliffs, N.J.: Prentice-Hall, Inc., 1969.

Index